# The Path to Autonomous Robots

Essays in Honor of George A. Bekey

T0137637

# The Path to Autonomous Robots

## Essays in Honor of George A. Bekey

Edited by

## Gaurav S. Sukhatme

 Springer

*Editor*
Gaurav S. Sukhatme
Department of Computer Science
University of Southern California MC 2905
Ronald Tutor Hall (RTH 405)
3710 South McClintock Avenue
Los Angeles, California 90089-2905
gaurav@usc.edu

ISBN: 978-1-4419-4675-1          e-ISBN: 978-0-387-85774-9
DOI: 10.1007/978-0-387-85774-9

springer.com

*To Professor George A. Bekey, on the occasion of his 80<sup>th</sup> birthday.*

# Preface

The principal chapters of this book form a collection of technical articles spanning many areas of research in robotics, these are followed by a set of short reminiscences and tributes written by former students of Professor George A. Bekey. Professor Bekey, a pioneer in robotics, retired from the University of Southern California (USC) in 2002 after serving on its faculty for forty years. He maintains an association with USC as University Professor Emeritus. Professor Bekey turned 80 in June 2008 - this is his Festschrift.

As one of Professor Bekey's former students, it has been my privilege to know him for many years. This book represents the collective warm feelings of his former students, who remember their association with him in the fondest terms.

Part I of this book is composed of technical chapters representing threads of active robotics research knitted loosely together. In many cases the themes of the chapters have their origins in the work the authors did when they were graduate students with Professor Bekey. These chapters are written for the reader interested in a sampling of modern research in Autonomous Robots. It is my hope that, for the serious reader, these chapters will serve as invitations to explore the field via further reading and research.

Part II of this book is composed of short, non-technical essays by former students of Professor Bekey. These serve to illustrate his multifaceted personality, and put on record the feelings of affection which many of us have for him.

Los Angeles                                                                 *Gaurav S. Sukhatme*
August 2008

# Contents

# List of Contributors

Arvin Agah
Electrical Engineering and Computer Science Department, University of Kansas, Lawrence, KS 66045, e-mail: agah@ku.edu

Eric L. Akers
Mathematics and Computer Science Department, Elizabeth City State University, Elizabeth City, NC 27909, e-mail: elakers@mail.ecsu.edu

Behcet Açıkmeşe
Jet Propulsion Laboratory, California Institute of Technology, Pasadena, CA 91109, e-mail: Behcet.Acikmese@jpl.nasa.gov

Dan Antonelli
Biokinesiology and Physical Therapy, Biomedical Engineering (retd.), University of Southern California, Los Angeles, CA 90089, e-mail: dantonel@usc.edu

Arun Bhadoria
Manufacturing Engineering Leader, Cummins Emission Solutions, Mineral Point, WI 53565, e-mail: arun.bhadoria@cummins.com

Douglas Brooks
Human-Automation Systems (HumAnS) Lab, Center for Healthcare Robotics@Health Systems Institute, Georgia Institute of Technology, Atlanta, GA 30332, e-mail: bdouglas8@gatech.edu@ece.gatech.edu

Charles de Granville
Symbiotic Computing Laboratory, School of Computer Science, University of Oklahoma, Norman, OK 73019, e-mail: chazz184@gmail.com

Willis G. Downing, Jr.
Professor of Biomedical and Electrical Engineering Emeritus, College of Engineering and Computer Science, California State University, Northridge, CA 91330, e-mail: wgdowning@netptc.net

Andrew H. Fagg

Symbiotic Computing Laboratory, School of Computer Science, University of Oklahoma, Norman, OK 73019, e-mail: fagg@cs.ou.edu

Christopher M. Gifford
Electrical Engineering and Computer Science Department, University of Kansas, Lawrence, KS 66045, e-mail: cgifford@eecs.ku.edu

Fred Y. Hadaegh
Jet Propulsion Laboratory, California Institute of Technology, Pasadena, CA 91109, e-mail: Fred.Y.Hadaegh@jpl.nasa.gov

Ayanna M. Howard
Human-Automation Systems (HumAnS) Lab, Center for Healthcare Robotics@Health Systems Institute, Georgia Institute of Technology, Atlanta, GA 30332, e-mail: ayanna@ece.gatech.edu

Gerard Jounghyun Kim
Digital Experience Laboratory, Korea University, e-mail: gjkim@korea.ac.kr

Theresa J. Klein
Department of Electrical and Computer Engineering, University of Arizona, Tucson, AZ, 85721, e-mail: tjk@ece.arizona.edu

M. Anthony Lewis
Department of Electrical and Computer Engineering, University of Arizona, Tucson, AZ, 85721, e-mail: malewis@ece.arizona.edu

Huan Liu
Department of Computer Science and Engineering, Arizona State University, Tempe AZ 85287-8809, e-mail: huanliu@asu.edu

Milan Mandić
Jet Propulsion Laboratory, California Institute of Technology, Pasadena, CA 91109, e-mail: Milan.Mandic@jpl.nasa.gov

Michael Merritt
Mission Systems, Northrup Grumman Corporation, e-mail: michael.merritt@ngc.com

L. Warren Morrison
Founder and Chief Technology Officer, ITG LABS LLC, e-mail: warrenmster@gmail.com

Chung Hyuk Park
Human-Automation Systems (HumAnS) Lab, Center for Healthcare Robotics@Health Systems Institute, Georgia Institute of Technology, Atlanta, GA 30332, e-mail: chunghyuk@ece.gatech.edu

Hae Won Park
Human-Automation Systems (HumAnS) Lab, Center for Healthcare Robotics@Health Systems Institute, Georgia Institute of Technology,

Atlanta, GA 30332, e-mail: hindol21@ece.gatech.edu

Robert Platt, Jr.,
Dexterous Robotics Laboratory, Johnson Space Center, NASA, Houston, TX,
e-mail: robert.platt-1@nasa.gov

Sekou Remy
Human-Automation Systems (HumAnS) Lab, Center for Healthcare
Robotics@Health Systems Institute, Georgia Institute of Technology,
Atlanta, GA 30332, e-mail: sekou@ece.gatech.edu

Daniel P. Scharf
Jet Propulsion Laboratory, California Institute of Technology, Pasadena, CA 91109,
e-mail: Daniel.P.Scharf@jpl.nasa.gov

H. Pete Schmid
Director, Advanced Technology (retd.), Raytheon Missile Systems Company,
Tucson, Arizona, e-mail: pete.schmid@alumni.usc.edu

Gurkirpal Singh
Jet Propulsion Laboratory, California Institute of Technology, Pasadena, CA 91109,
e-mail: Gurkipal.Singh@jpl.nasa.gov

Joshua Southerland
Symbiotic Computing Laboratory, School of Computer Science, University of
Oklahoma, Norman, OK 73019,
e-mail: Joshua.B.Southerland-1@ou.edu

Richard S. Stansbury
Department of Computer and Software Engineering, Embry-Riddle Aeronautical
University, Daytona Beach, FL 32114,
e-mail: richard.stansbury@erau.edu

Gaurav S. Sukhatme
Department of Computer Science, University of Southern California, Los Angeles,
CA 90089-0781, e-mail: gaurav@usc.edu

Monte Ung
Department of Electrical Engineering, University of Southern California, Los
Angeles, CA 90089, e-mail: ung@ceng.usc.edu

Di Wang
Symbiotic Computing Laboratory, School of Computer Science, University of
Oklahoma, Norman, OK 73019, e-mail: di@cs.ou.edu

Dit-Yan Yeung
Department of Computer Science and Engineering, Hong Kong University of
Science and Technology, Clear Water Bay, Hong Kong, China,
e-mail: dyyeung@cse.ust.hk

Bin Zhang

Department of Computer Science, University of Southern California, Los Angeles, CA 90089-0781, e-mail: binz@usc.edu

Yu Zhang
Department of Computer Science and Engineering, Hong Kong University of Science and Technology, Clear Water Bay, Hong Kong, China, e-mail: zhangyu@cse.ust.hk

# Part I
# Recent Research in Autonomous Robots

The eight chapters which make up this part of the book span separate research areas in Autonomous Robots. It has long been clear that robots will play a significant role in applications which are dirty, dull or dangerous. However many fundamental challenges remain to be addressed before robots become ubiquitous. The chapters in part I of this book address some of these challenges. Some chapters take a basic research perspective, and others take an application-driven viewpoint. In the first chapter, Agah and colleagues at the University of Kansas describe the design of mobile robots for seismic and radar remote sensing of ice sheets in polar regions - one of the most hostile environments on the planet. This is followed by an account of autonomous spacecraft from Hadaegh and colleagues at the Jet Propulsion Laboratory, another example of robots in extreme environments. In chapter three Sukhatme and Zhang at the University of Southern California describe the design of networked robotic systems with applications to sensing and sampling the aquatic environment. While the first three chapters deal largely with mobile robots (albeit with dramatically different kinds of mobility), clearly autonomy implies the ability not only to sense and move, but to manipulate the environment. In chapter four, Fagg and colleagues at Oklahoma University describe their recent research in robotic grasping using inputs from vision systems. Designers of autonomous systems cannot help but admire the plethora of natural autonomous systems in the world around us - living beings. In chapter five, Howard and colleagues at the Georgia Institute of Technology discuss advances in both manipulation and mobility which are inspired by human control systems. This is followed by a discussion by Kim on new directions in human-robot interaction. Lewis and Klein also take a biomimetic approach to autonomous robot design, and provide an introduction to neurobotics. A fundamental characteristic of living beings is that they learn. In this spirit, the eighth and final chapter by Yeung and Zhang investigates the feasibility of applying multi-task learning to the problem of inverse dynamics.

# Chapter 1
# Mobile Robots for Polar Remote Sensing

Christopher M. Gifford, Eric L. Akers, Richard S. Stansbury, and Arvin Agah

**Abstract** Mobile robots are becoming more heavily used in environments where human involvement is limited, impossible, or dangerous. These robots perform some of the more dangerous and laborious human tasks on Earth and throughout the solar system, many times with greater efficiency and accuracy, saving both time and resources. As we explore further away from Earth, higher levels of autonomy are also becoming more desired in such applications, one of them being remote sensing. This chapter covers mobile robots that have been designed and built at the University of Kansas to facilitate seismic and radar remote sensing of ice sheets in polar regions. These robots have been developed for and deployed in unstructured, polar environments. System designs, components, deployment and data acquisition algorithms, and experimental results are discussed. In this chapter, future applications, such as an autonomous multi-robot seismic surveying team, are simulated. Future planetary missions will hopefully incorporate similar robotic systems to conduct in-situ experiments on other planets.

Christopher M. Gifford
Electrical Engineering and Computer Science Department, University of Kansas, Lawrence, KS 66045, e-mail: cgifford@eecs.ku.edu

Eric L. Akers
Mathematics and Computer Science Department, Elizabeth City State University, Elizabeth City, NC 27909, e-mail: elakers@mail.ecsu.edu

Richard S. Stansbury
Department of Computer and Software Engineering, Embry-Riddle Aeronautical University, Daytona Beach, FL 32114, e-mail: richard.stansbury@erau.edu

Arvin Agah
Electrical Engineering and Computer Science Department, University of Kansas, Lawrence, KS 66045, e-mail: agah@ku.edu

## 1.1 Introduction

At the University of Kansas, the Center for Remote Sensing of Ice Sheets (CReSIS) (48) performs polar research to gather data and model ice sheets to better understand climate change and its effects on sea level rise. We have designed, built, and deployed mobile robots to autonomously traverse polar terrain and support experiments in Greenland and Antarctica. The problem we are faced with is to facilitate the use, and increase the efficiency of seismic and radar data acquisition in these types of environments. Integration of autonomy and mobility into remote sensing methods can potentially improve and enhance the process.

One of the sensors used to perform this research is a seismic sensor, or geophone. These highly sensitive geophones detect vibrations in the ground which can be recorded as images. These images, for example, can show characteristics of the subsurface, detect cracking (fault) locations, as well as provide information on what is beneath the ice sheets. CReSIS has also developed ground and ice penetrating radars that can also provide highly detailed images of the subsurface. We have used these radars to identify layers and areas of potential water within Earth's ice sheets, as well as the extent of water presence at the interface between the ice sheet and the underlying bedrock. Together, these remote sensing techniques allow us to study and further understand how polar regions are changing in response to climate change.

Research in the field of robotics has been focusing on accurate sensing and autonomy, mostly performed in structured environments such as factories and homes. Robotic applications involving remote sensing and in-situ experiments in unstructured environments have, however, been limited. Not only are navigation and actuation in such environments difficult problems, autonomy in hazardous environments represent larger challenges (166).

Another important aspect of integrating robotics and remote sensing surveys is that it limits human involvement, the most dangerous and costly portion of a survey. The accuracy and quality of acquired data can also be potentially increased. For polar and planetary environments, this becomes extremely important for safety, reliability, and resource consumption. Furthermore, robotics increases precision and introduces repeatability into an otherwise time-consuming and complex human task. In particular, because seismic deployment is labor-intensive, expensive in terms of time and cost, and possibly dangerous, autonomously performing such tasks using mobile robots can be highly beneficial. Therefore, the goal is to combine robotics research with remote sensing systems to autonomously image the subsurface for polar and planetary applications.

As we explore further away from Earth, a need for a coupled increase in intelligence and autonomy arises as a result of large communication delays between robots and their remote human operators. During communication delays, robots could navigate to new locations, perform experiments, or build a map of an unexplored region of the surface. Other planets and their moons offer various unique and challenging environments for ground-based robots, ranging from the extreme heat and pressure on the surface of Venus to the icy landscape of Jupiter's moon Europa. Future missions to the icy moon Europa may one day include a multi-robot seismic mission to learn more about its subsurface and the ever-present search for Earth-like processes and life in the solar system.

This chapter covers mobile robots that have been designed at the University of Kansas to facilitate seismic and radar remote sensing in polar regions. Section 2 introduces two polar mobile robots, challenges and survivability issues, and their field operations in Greenland and Antarctica. Section 3 discusses designs and simulations of single- and multi-robot approaches to automated seismic surveying. Finally, Section 4 discusses the conclusions and future directions of work related to these systems.

## 1.2 Polar Mobile Robots

At the University of Kansas, mobile robots have been utilized to autonomously support the Center for Remote Sensing of Ice Sheets (CReSIS) (48) and Polar Radar for Ice Sheet Measurements (PRISM) (100) projects. Mobile robots are required to transport a radar system across the ice sheet, following a precise movement pattern and tow an antenna, while ensuring the safety of the radar system. Mobile robots designed for harsh environments such as the polar regions can increase the efficiency of data collection in field experiments, reduce the potential dangers to humans, and enable experiments beyond the regular field seasons. When the data collection process requires several days, weeks, or months, special considerations must be made for the survivability of these robots. Two mobile robots, shown in Figure 1.1, have been evaluated in Greenland and Antarctica during three field seasons for autonomous data acquisition using radar remote sensing. The following sections discuss challenges and survivability issues associated with polar robotics, platform components, sensing, and field operations.

**Fig. 1.1** MARVIN I (left, in Greenland) and MARVIN II (right, in Antarctica) equipped with components for polar operations.

### 1.2.1 Challenges and Survivability Issues for Polar Robotics

There are many challenges faced by polar robots, as described in detail in (11). Polar robots must operate in extremely cold temperatures. The temperatures at the Amundsen-Scott South Pole Base have been recorded to vary from $-13.6^o$ Celsius to $-82.8^o$ Celsius (118). To address short-term survival during polar Summer seasons, onboard robot components must at least meet a minimum operating temperature of $-40^o$ Celsius. High wind speeds also produce many unique challenges to polar survival. All components of the mobile robot must be ruggedized to handle high wind speeds. At the South Pole Station, the average wind speed is 5.5 meters per second with a maximum recorded gust of 24 meters per second (118). Polar winds erode the ice sheet creating surface obstacles such as snow drifts and sastrugis. The blowing snow further contributes to low visibility.

Obstacles such as sastrugis can cause damage to the mobile robot. Sastrugis are dune-like ice structures that result from erosion of the ice sheet's surface and may have only one steep side. Due to their unique shape, blowing snow, a lack of surface contrast, and inadequacies of stereo vision, sastrugis are difficult to detect. A robot cannot remain stationary for an extended period of time without snow drifting along one side. If a robot must suspend operation temporarily during a snow storm (i.e., lack of solar energy), it must maneuver such that it will not become stuck.

Any venture away from an established camp poses the risk of encountering a crevasse. Crevasses may be hidden by a layer, or "canopy", of snow. To date, airborne radar has been the most used mechanism for locating crevasses within an area, but that may be insufficient for long distance traversals. Researchers have utilized ground penetrating radar mounted on a boom several meters ahead of the vehicle to detect crevasses, which requires that the vehicle travels quite slowly (15). Smaller robots may not be capable of supporting such a configuration because of the mass required to support the antenna boom. Additionally, vision sensors are often insufficient for operation within polar environments. Blowing snow may obscure cameras, sonar, and laser measurement sensors. Depth perception is also limited when all backgrounds are white. Cloudy days may reduce visibility due to a lack of contrast. Currently, only millimeter-wave radar is capable of operating in these low visibility situations, but those are often heavy, consume power resources, and could be infeasible in terms of cost. Figure 1.2 presents examples of limited visibility and snow drifting that was formed overnight in the polar regions.

### 1.2.2 MARVIN I

The first mobile robot for the PRISM project, MARVIN I, was constructed using a MaxATV Buffalo (141) All Terrain Vehicle (ATV) as the mobile platform. This section presents a description of the robot, and more thorough descriptions are available in (10; 81). The six-wheeled MaxATV, shown in Figure 1.3, features a skid-steering drive system. The robot has a ground clearance of about 12.5 cm. It can tow up to

**Fig. 1.2** A snow drift that formed overnight behind MARVIN II while in camp (left), and limited visibility due to blowing snow (right).

454 kg and haul 408 kg. A custom enclosure, shown in Figure 1.1, was built on top of the rover to house the robot's sensors, actuators, computers, and all the radar equipment. A gas generator provides power to the onboard systems.

**Fig. 1.3** Original Buffalo Max ATV used as the base platform for the MARVIN I mobile robot.

Linear actuators are used to automate the platform. The left and right brakes are controlled using linear actuators attached to the respective control handle for each. The engine's throttle is controlled using a linear actuator, attached to the rover's throttle control cable. The actuators are magnetically driven and provide 5 mm position accuracy. Localization was handled using a Topcon real-time kinematic (RTK) GPS (179) for centimeter-level accuracy (required for radar measurements), along with a gyroscope and inclinometer to provide the robot's orientation. A compass can not be used in polar robots due to the skew caused by the close proximity to the magnetic poles. A laser range-finder is used to detect obstacles. An internal temperature sensor provides feedback regarding the current temperature within the environmental enclosure. Lastly, a weather station and a pan-tilt-zoom camera are included for outreach activities of the project. Figure 1.4 presents MARVIN I and its components in Greenland (2004), and also shows the mounting of a radar antenna on the platform. Automated navigation is handled using a ruggedized laptop.

**Fig. 1.4** MARVIN I in Greenland (2004) with external sensor components. A radar antenna mounted for data acquisition is also shown.

When designing and building MARVIN I, many design decisions such as electric versus gas powered engine, the type of vehicle, tracked versus wheeled, etc. had to be made without the ability to test in the same conditions in which it had to run. As a result, a virtual prototype of MARVIN I was created and tested in simulation. MSC.visualNastran 4D (116) was used for both the modeling and simulation of the vehicle. Figure 1.5 shows the model that was created for performing platform and towing tests. The model does not perfectly match the actual vehicle, however, as there are many shapes and contours that are difficult and unnecessary to model. Also, the weight distribution of the vehicle was not completely known. The weight of the vehicle was evenly distributed around the base of the vehicle when the exact weight was known. The model for the antenna was created simply as a flat rectangular box, also shown in Figure 1.5.

The simulation experiments were designed to answer some specific questions about rover performance. First, what should the starting point be for the safety parameters? The safety parameters included the maximum slope the rover could climb (the pitch angle) and the point at which the rover would roll over (the roll angle), and how would the rover handle while performing basic movements and with different load configurations? The results from these tests are described in (9), along with a full description of the modeling and testing process. Software and control will be discussed in the next section, as it was inherited by the second version of the robot: MARVIN II.

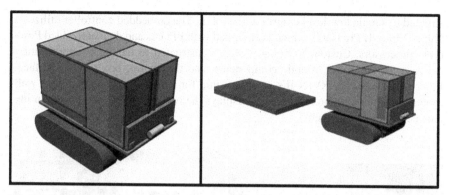

**Fig. 1.5** Tracked vehicle model used for each simulation experiment (left), and the model towing an antenna sled.

## 1.2.3 MARVIN II

In order to address some of the limitations of MARVIN I (12; 81), MARVIN II, shown in Figure 1.1, was designed and built. Many of the decisions made regarding the design of the new robot were traceable to the lessons learned from MARVIN I (12). MARVIN II was built using a RangeRunner All Terrain Vehicle (ATV) from TerraTrack (174) as the mobile platform base vehicle. The new vehicle has a 22.9 centimeters ground clearance, a 32 horsepower diesel engine, and a hydrostatic differential drive system. Hydraulic motors with built-in speed sensors allow for a finer control. Turns can be produced by speeding up one side, slowing down one side, or both. This improves the system in two ways: (1) horsepower is not lost to braking, and (2) one track no longer has to be dragged in order to turn. Another difference is that the new base vehicle is built with tracks whereas the first is a six-wheeled vehicle with a track kit. The tracks on MARVIN I had a tendency to detach when turning on hard surfaces.

The sensor suite of MARVIN II is quite similar to its predecessor. The weather station has been eliminated in place of a ground-based solution. The camera no longer pans and tilts, but rather has a fixed mount because the motors could not handle the vibration caused by a moving vehicle on the ice sheet. Speed sensors have been integrated with the hydraulic motors used to drive the vehicle. Each speed sensor provides a fixed number of pulses per revolution, which is used to determine the speed of each track. During early development of MARVIN II, a Garmin (67) handheld GPS receiver was used in place of the Topcon GPS. The Topcon GPS receiver was later utilized, but no longer in RTK differential mode due to the longer paths required during the 2005/2006 Antarctic field season.

A control box has been constructed to house an embedded control system, internal robotic sensors, and the power supplies for the various subsystems. Toggle switches allow the various subsystems to be powered individually. I/O ports such as USB, RS232, and Ethernet are provided such that multiple computers and external devices may be integrated with the control box. An embedded controller was con-

structed to handle low-level control of the robot. The embedded controller utilizes a
Xilinx Virtex-II Pro (197), which is equipped with FPGAs and dual embedded Pow-
erPC processors. Custom I/O boards were constructed to handle communication
between the controller, robotic components, and the control box's user interface.
Figure 1.6 shows MARVIN II's actuator setup for autonomous navigation as well
as the custom control box. Figure 1.7 displays the robot's capabilities for mounting
data acquisition equipment.

**Fig. 1.6** MARVIN II's actuator setup for autonomous navigation (left), and custom control box
housing the embedded controller and other components (right).

**Fig. 1.7** MARVIN II platform with environmental enclosure, radar crate mounted on the front, and
additional fuel on the back (left); and larger crates mounted on both sides of the custom aluminum
frame (right).

The control system is divided between the embedded controller to control the speed of the tracks and a laptop that communicates with the embedded controller. The laptop determines where and at what speed the vehicle should move. The laptop uses a wireless network to communicate remotely. Unlike MARVIN I, the new control software is written such that the two levels of control are decomposed from the original monolithic control system. At the highest layer, the list of waypoints is evaluated and the waypoint navigator is given a new path to traverse. At the waypoint navigator layer, the sensors are monitored to determine and assign desired speeds for each track in order to drive the robot toward the path. Finally, at the lowest level, the embedded controller adjusts the actuator state until the desired speed is reached. This modularity allows the laptop to control the higher level actions of the robot such as setting the desired speed of each track, performing obstacle avoidance, and making route corrections. The embedded controller handles the lower-level control which attempts to keep each track moving at the requested speed while adjusting for bumps, hills, or anything that might hinder the speed of the tracks.

## 1.2.4 Software Architecture

A software application programming interface (API) was developed for PRISM and CReSIS mobile robotics research. The API was designed with an emphasis on portability, robustness, and maintainability. As the underlying hardware of the mobile robot changes, the overall structure of the software system should remain intact and only require minor changes to adapt to the new hardware. The software libraries should also be general enough so that they could be utilized for a variety of mobile robots.

Figure 1.8 presents the robot API. Devices are modeled as either sensors or actuators, and several abstract classes for each exist. When implementing for a particular robot, the abstract sensor classes are implemented as drivers for the particular component. By using these abstract classes, it is possible to change the robotic platform while maintaining the same control system and automation software, utilizing the same intelligence algorithms. Figure 1.9 shows the mapping from each platform to a generic control system type. Figure 1.10 demonstrates the execution of software to guide a robot to move in an "S-pattern" on both a Nomadic Scout testbed and the MARVIN I polar rover.

The primary advantage of creating a reusable architecture for robot software development is to simplify testing the robot. The same algorithms may be ported between simulation, testbeds, and the field robots. When operating large mobile robots, small tests to prove control logic can become quite logistically challenging without sufficient use of prototypes, and this software helped bridge the necessary gap to do so. The same challenge holds for robots for polar regions.

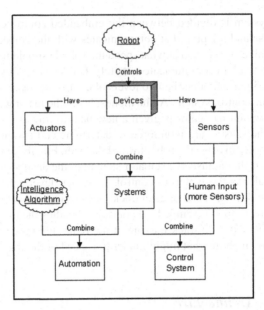

**Fig. 1.8** Robot API for PRISM/CReSIS mobile robots.

## 1.2.5 North Greenland Ice Core Project (GRIP) Camp Operations

Research during the Greenland 2003 field season focused on testing each of the individual components of MARVIN I. Components were subjected to a deep freeze upon arrival on the ice sheet, and each was tested to ensure that it could reliably operate within the polar conditions. The rover was tested over several long-range traversals to determine how it handled when maneuvering on the snow. Figure 1.11 shows MARVIN I towing a prototype radar sled across the ice sheet. MARVIN I became stuck in snow only once. During the North GRIP camp (121) experiments, the sensors proved reliable. Figure 1.11 also shows the laser range-finder being tested by imaging a man-made snow wall. Other experiments included attempting to detect a simulated crevasse with the laser range-finder, which failed to yield positive results. Evaluation of the GPS receiver was performed to determine sufficient satellite coverage. From the results, it was clear that the minimum four necessary GPS satellites were available at all times.

## 1.2.6 Summit Camp Operations

After a year of work, MARVIN I was sent back to Greenland in the Summer of 2004 to test its automation. It was integrated with a radar system such that it would autonomously move across the ice sheet following a precise path so that radar data

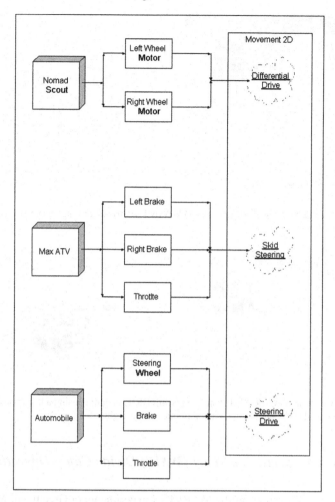

**Fig. 1.9** Mapping of platform to actuator API class to control systems.

could be collected. Unlike the previous experience at North GRIP, it was much warmer at the Summit Camp in 2004. As a result, the rover became stuck periodically. This happened most frequently during turns in which one of the tracks was stopped. The camera's pan-tilt motor also failed due to excessive vibrations. Ultimately, MARVIN I's transmission failed due to the payload being too heavy, and possibly a buildup of ice on the robot's axles. These were lessons learned that were then applied to the MARVIN II polar robot design.

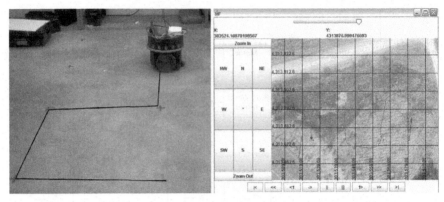

**Fig. 1.10** Nomadic Scout testbed (left) and MARVIN I (right) performing "S-pattern" using identical control software.

**Fig. 1.11** Evaluation of MARVIN I systems at North GRIP, including using a laser range-finder to image a snow wall (left) and towing a mockup radar array (right).

## *1.2.7 West Antarctic Ice Sheet (WAIS) Divide Camp Operations*

Having learned lessons from the MARVIN I experiments in Greenland, MARVIN II was deployed in Antarctica in 2006 to support mono-static and bi-static radar applications. Figure 1.12 shows the MARVIN II rover towing two different radar systems in the field. MARVIN II is the current ground-based robot used by CReSIS for polar field expeditions.

Figure 1.13 illustrates the deployment of the two robots in the field. The left figure plots the movement of MARVIN I between two waypoints with 25 meter grid spacing. Despite a high-end GPS providing centimeter-level accuracy, MARVIN I could not be mechanically controlled to provide a sufficiently straight path for autonomous radar survey. This demonstrates the impact of its transmission failure. The middle figure plots MARVIN II traversing between two waypoints with and without high-end GPS enabled with a 5 meter grid separation. Without RTK GPS, MARVIN II oscillated about the line. With the high-end GPS enabled, the amplitude of oscillation about the line was dramatically reduced, as shown in Figure 1.13. The

**Fig. 1.12** MARVIN II rover towing a radar sled for a mono-static application (left) and a radar sled for a bi-static application (right) in Antarctica.

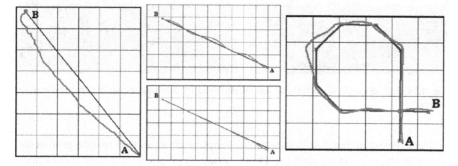

**Fig. 1.13** Navigation results for MARVIN I with 25 meter resolution (left), MARVIN II straight line with 5 meter resolution (middle), and MARVIN II turn patterns with 5 meter resolution (right).

right image in Figure 1.13 plots (with 5 meter grid spacing) MARVIN II's path as it performs a turn maneuver to switch between paths.

## 1.3 Robotics-Based Approaches to Seismic Surveying

Seismic sensors, also known as geophones, are sensors which transfer vibration waves as a series of analog signals based on the composition of the material beneath the surface and the travel times of the measured seismic waves. Widely used for oil exploration, geophones are activated by a seismic source, which can range from striking the ground to a very large explosion or a natural geologic event. The source sends elastic vibration energy down into and through the subsurface so as to eventually reflect and refract back after interaction with the internal layers. Based on the travel times, wave velocities, and the received signals from a series of geophones, Seismologists can digitize, filter, and analyze the results to learn such facts as water table depth, fault location, and rock layer boundaries. The refracted and reflected paths can provide structural information of the subsurface (136). The physical properties of the rocks and beds affect travel times of seismic waves. These travel times, along with the waveform and spectra, are then used to deduce information about the subsurface and internal layering. Various styles and models of seismic sensors exist for many applications on land, snow, and at sea. A single geophone and linear array of deployed geophones are shown in Figure 1.14.

**Fig. 1.14** Spiked geophone (left), and several deployed by inserting the spikes into the surface (right).

Deployment of geophones translates into how each is inserted into the ground (or alternatively, rests on the surface). During manual operations involving tens or hundreds of geophones, each is typically stepped on or hand-pressed into the ground. If necessary, holes are dug prior to deployment to create a shelter for the sensor to record its data. This is a timely and arduous task, especially when performed in a polar environment. Coupling of the geophones with the ground directly affects the data quality and frequencies that can be recorded. Geophones must also be arranged in a centimeter-level precision grid of equal spacing while being oriented no more than $10^o$ from the Earth's gravitational vertical.

Seismic arrays can be formed to acquire a map of the subsurface, allowing detailed imaging at many different resolutions and depths. Higher frequencies and close (sub-meter to tens of meters) spacing results in a highly detailed, shallow image of the subsurface. Deeper imaging requires sparse deployment and long distances from a powerful source, with spacing ranging from hundreds to thousands of meters. Furthermore, high frequency acquisition translates into more accurate images. If geophones are positioned in a straight line, a seismic survey will result in a two-dimensional (2D) image/slice of the subsurface. If the geophones are aligned in a square or rectangular grid pattern, a three-dimensional (3D) view of subsurface characteristics will result. A fourth dimension, namely time, can be introduced to image movement of the subsurface and materials.

## 1.3.1 Related Work

Very little work involving robotic deployment and retrieval of seismic sensors has been done to date. However, work done in regular environments can provide helpful information. As stated in (84), the future of seismic surveying on land is the elimination of cables. Seismic networks can also be of less weight and easily scalable in terms of network size, structure, and shape.

The University of Kansas Geology Department recently developed an "autojuggie" (180) capable of planting 72 geophones in 2 seconds using a hydraulic press and structured array system. Several variations of the autojuggie have also been developed and field-tested (165). These variations include automated deployment using farm equipment and deployment of closely-spaced lines of seismic sensors for ultra-shallow imaging. Structures were used to simultaneously press all sensors into the ground, and simultaneously retrieve all geophones when finished. Care was also taken to try to reduce crosstalk between sensors through the deployment structures. These approaches are still human-operated in that they use existing farm equipment and vehicles as a means for deployment and retrieval. Scalability and robustness of this approach is limited as well.

Land streamers are a method inherited from the marine seismic community, which deploy a series of geophones by dragging them along the surface. Acquisition takes place when stopped, where all geophones typically rest on metal plates rather than being physically inserted into the ground. This increases deployment efficiency by reducing the time required for insertion and orientation of the sensors, as well as reducing transportation time from one site to another. In (164), multiple land streamers were pulled alongside each other at the same time using an ATV. Individual land streamers were spaced equidistant from one another to a towing structure so as to create a wider 2D array. Results were acceptable for relaxed seismic requirements, but would not be applicable under higher frequency situations. Other efforts have also been published (60; 95; 155) that employed single streamers in a polar setting, or specifically designed for shallow data acquisition (181; 182). Survey requirements and weather conditions dictated the geophone spacing, streamer length, and materials used to construct the streamers. Several streamer designs have been attempted in these works, ranging from the 1970's to the present. The Kansas Geological Survey made their land streamer more rugged by encasing it and all wiring in a fire hose (92).

NASA, working with Georgia Tech and Metrica, Inc., developed an Extra-Vehicular Activity Robotic Assistant (39) capable of being handed a geophone and inserting it into soil using a seven degree-of-freedom manipulator and a three-fingered gripper (113). The 4-wheeled mobile robot could not perform the full deployment task and was not made to retrieve the planted geophones. The main purpose of this robot was to assist activity-suited humans in the field by performing some tasks on its own. A trailer containing the geophones was pulled so the human could hand them to the robot or store other various supplies.

## *1.3.2 Robotics-Based Approaches*

Many possible mechanisms exist for deployment and retrieval of seismic sensors (71; 72). Automating the process makes detailed imaging on many scales much more feasible.

### Individual Deployment

Individual deployment covers those methods that deploy and retrieve a single geophone at a time. This mechanism could be a robotic arm, crane-like apparatus, air-powered device, planter, or any other form of pick-and-place device. In many planting, weeding, and picking projects, robotic manipulators are utilized to help automate the process. This type of mechanism is responsible for pressing into, orienting, and pulling from the ground all seismic sensors and placing them in a transport area, charging station, or organized rack. Size, shape, and weight influence the overall design of platform(s) performing the task. Issues with this category involve orientation, positioning, and weather. Autonomously dealing with and keeping track of the tangling maze of seismic cable also represents a formidable challenge. A positive aspect of this approach is the millimeter repeatability and precision that manipulators offer. However, finding and retrieving geophones, manipulator payload, required pushing power, and gripping the geophone are all major difficulties inherent to this approach.

### Array Deployment

The array deployment involves an array to deploy and retrieve a set of geophones, their cables, and all necessary storage equipment. Seismic sensors would be pre-set into the array, taken to the field location, and simultaneously (e.g., hydraulically) pressed into the ground at equal spacing, tilt, and elevation. When ready for retrieval, the structure is raised to remove the sensors from the ground. Multiple arrays could be pieced together to record larger areas. The design could also permit variable sensor spacing to perform different resolution imaging. Geophone spacing, orientation, and deployment depth are therefore controlled for the entire seismic array. Scalability in terms of size and imaging resolution is lacking due to being pre-built, however, and this approach is also still wired.

### Land Streamers

The idea of land streamers came from marine seismic surveying, which involves constantly towing marine streamers under water with use of pulse guns for the sound source. Land streamers are a non-insertion seismic method where geophones are wired in series and towed on the surface to acquire seismic data. When the recording location is reached, the towing vehicle stops so that seismic acquisition can take place. One or more of these streamers can be towed in parallel to cover larger areas and perform 2D or 3D imaging.

Autonomous seismic acquisition can then be accomplished with, for example, the MARVIN II robot. The Webots (51) simulation environment was used to test GPS waypoint navigation, driving, and turning algorithms, whereas MSC visualNastran (116) was employed to simulate pulling and drag abilities of our autonomous rovers. Figure 1.15 shows a MSC visualNastran simulation involving three streamers, where each box represents an enclosed geophone.

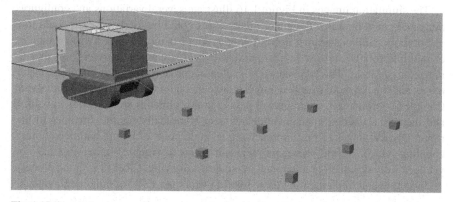

**Fig. 1.15** Simulation image of an autonomous robot towing a three-streamer array, used for studying towing of streamers and how turning affects strain of the towing structure and travel of the streamer components.

This mechanism could extend to cover a long distance behind the rover as well as widen coverage width by using multiple streamers. An attractive aspect of this category is the ability to choose and change spacing of sensors within and between streamer lines. Other advantages of this approach are its ease of transport, efficiency, simplicity, and no need for geophone insertion. The main advantage to these types of systems is speed and the amount of seismic data that can be recorded with fewer personnel. The unattractive characteristic of this approach is its lower coupling. This may cause the geophones to miss higher frequencies, resulting in less detailed seismic images. Some research has shown that, in some environments, performance between conventional geophones and land streamers are very similar.

**Hybrid Streamers**

It has been proposed that a hybrid combination of land streamers with increased coupling would be a good alternative (71). There are several design options to increase hybrid streamer coupling:

- Employ a trenching or plowing attachment to prepare the ground to drag the streamers below the surface for wind protection and to rest flat for orientation purposes;
- Add weight to each streamer node;
- Change plate size and/or geometry;
- Increase the surface area the plates have with the ground;
- Heat streamer plates for snowy/polar environments so the melt can refreeze to ice, giving a more rigid surface contact for the plates; and
- Drill the geophone into the ground like a threaded screw.

Accordingly, a furrowing, plowing, or trenching apparatus could be attached to a mobile robot. The robot would power all equipment, have seismographs onboard for seismic data conversion and storage, and have a data cable which would act as both the data transmission and communication medium for the entire system.

The simulation images in Figure 1.16 illustrate several variations and configurations that could be utilized. One or more robots could be used and each could tow one or more parallel streamer lines. A single robot can tow a single hybrid streamer, or multiple robots can tow several parallel hybrid streamers and work together to image larger areas. The advantages of such an approach are better coupling, faster travel, and the potential to collect much more data with far fewer personnel involved. Complex coordination, communication, and node collisions are essentially avoided and there is no added attachment/detachment complexity for the streamer to the robot. Disadvantages to this approach are a single point of failure and overcoming coupling issues. Hybrid streamers represent a new seismic technique that has not yet been fully designed or attempted in modern surveying. CReSIS is in the process of designing and implementing these hybrid techniques (74) for polar deployment.

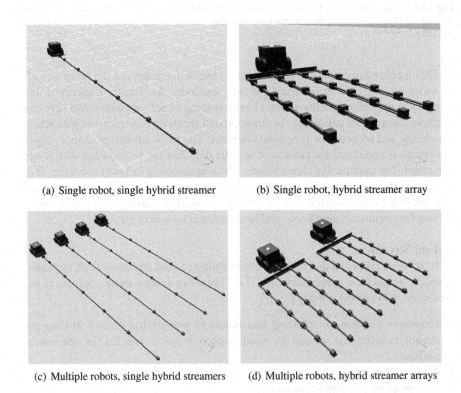

(a) Single robot, single hybrid streamer          (b) Single robot, hybrid streamer array

(c) Multiple robots, single hybrid streamers     (d) Multiple robots, hybrid streamer arrays

**Fig. 1.16** Simulation images illustrating variations of hybrid streamers towed by mobile robots.

## Multi-Robot Seismic Surveying Team

Based on the demonstrated success of multi-robot systems (distributed robotics) (8; 28), we have proposed use of a multi-robot seismic surveying team. The multi-robot seismic surveying approach involves a team of several autonomous, mobile robots that are smaller in size to deploy geophones and traverse the environment. They work together to precisely align into a seismic grid pattern. Each robot represents a mobile node that deploys and retrieves its own geophone. Power is provided by onboard sources, where each robot contains the necessary digitizing, storage, and communication hardware for seismic acquisition.

A mobile robot can inject into or place a geophone onto the ground while protecting the deployed sensors from the wind and weather using an environmental enclosure. Team size can be relatively small, such as a 25-robot team forming a 5x5 seismic grid, or extremely large, consisting of potentially hundreds of robots forming grids of any size, shape, and spacing for different seismic resolution applications.

There are various ways that the team could move into position. Robots could move one at a time in a certain fashion, by rows or columns, or dynamically align while all moving at once. Positioning one robot at a time takes longer, but could help increase accuracy and reduce collisions (73). Dynamically forming the seismic grid would take less time and would likely be a more flexible solution, but would suffer from inherently being less precise.

Figure 1.17 illustrates a shape formation scenario in simulation. The robots coordinate which GPS positions they travel to based on a desired grid shape and spacing, as provided by the main robot.

**Fig. 1.17** Simulation images showing a team of mobile robots forming a square seismic grid, one-by-one from top-right to bottom-left.

The advantages of the multi-robot seismic sensor network approach are that it would be faster than a human team for large arrays, removes cumbersome wires from the system, and allows safe remote sensing while being able to dynamically adjust to the environment. This distributed methodology removes the single point of failure. This is also a new seismic method that has not yet been attempted, mainly because it remains too challenging at this time. The main bottlenecks lie in highly precise alignment of a team of mobile robots at any scale and any environment (73), along with properly aggregating the seismic data. This is the most desirable approach based on its mobility and ability to image at any resolution, shape, and scale. This might also provide faster network assembly, especially for a large and remote team. Dropping the robot team from an aerial vehicle to assemble, record, and perform multiple missions represents a futuristic option in this category. A design has been proposed for such a mobile robot team platform, as well as precise grid formation schemes such that the team could form a precise seismic grid one at a time or in a dynamic fashion (71; 73). This category of seismic sensing has not been formally performed, but is currently being studied at CReSIS.

### Seismic TETwalker Mobile Robot

The NASA/Goddard Space Flight Center in Maryland is developing a new approach for the sustainable and affordable robotic exploration of the solar system: the Autonomous Nano-Technology Swarm (ANTS) (49). The basic unit of the structure is a tetrahedron consisting of nodes interconnected with struts that can be reversibly and/or partially deployed or stowed to allow forward motion on a surface at a controllable scale or gait. 3D networks are formed from interconnecting reconfigurable tetrahedra, making structures which are scalable, massively parallel systems. These robots have been thusly named Tetrahedral walkers, or TETwalkers.

With these aspects in mind, the 4-TETwalker's center node that it uses to shift its center of gravity has been studied for geophone deployment (41; 145; 189). As the robot could potentially deploy through any "face" of its tetrahedral structure after toppling, a center node was designed for this purpose with angled faces parallel to the faces of the TETwalker's outer shape. The corresponding vertical strut is used to push/insert the sensor into or onto the surface. Lengthening and shortening struts allow the deployment node to precisely place the geophone, as well as vertically orient it for proper data acquisition on tilted surfaces.

The design utilized the 4-TETwalker's center node to house one or more geophones, each with a dedicated spike or surface plate. The robot could then place the center node down onto the ground by extending the strut that is vertical during deployment. Figure 1.18 shows models of the center node, designed with a geophone and spike for downward deployment using any side of the 4-TETwalker.

**Fig. 1.18** Center node design with geophones and spikes (far left image), and the upright 4-TETwalker vertically aligning (second image from left) and deploying the downward-pointing geophone in a gimbaled manner (right two images).

Deployment and mobility simulations were conducted to study the dynamics of 4-TETwalker motion and strut extension properties for toppling. The struts are modeled as lightweight actuators capable of extending and retracting given the weight of the system. Figures 1.19 shows models used for deployment and mobility simulations. Further simulation studies are taking place to improve these models.

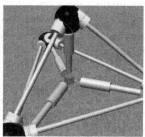

**Fig. 1.19** Simulation images of models used to study the physical characteristics of center node deployment and retrieval.

## 1.4 Conclusion

This chapter described the research efforts at the University of Kansas and CReSIS, focused on developing single- and multi-robot systems for remote sensing and exploration applications in polar environments, with extensions possible to planetary missions. These systems represent established building blocks for further research in the areas of robotics-based seismic surveying and related robotic deployment designs, and mobile robots for autonomous data acquisition in extreme polar settings. Using such robotic systems can decrease human involvement and reduce risk of science platforms in unexplored environments.

Ground-based remote sensing experiments have been extremely limited in planetary robotics due to the combination of lower reliability, increased risk, and limited technical readiness levels. As we explore further from Earth, increased autonomy is desired to compensate for the downtime associated with communication delays. We envision autonomous mobile seismic and radar mapping networks on other planets, such as Jupiter's icy moon Europa, to explore their subsurface characteristics. Information that can be acquired from such experiments could provide insight into the developmental history of other planets, and also aid in the quest for better understanding of the solar system.

As part of this work, novel designs and algorithm simulations demonstrate that a multi-robot seismic surveying team represents a futuristic approach for remote sensing on Earth and other planets. We have developed a low-cost multi-robot exploration and mapping team that will be used to extend this task from simulation to reality. We also plan to scale the system up by building more robots, further research low-cost sensing systems, and introduce more sophisticated multi-robot coordination and exploration techniques. We would also like to integrate more formal science and sensor components so as to perform distributed science experiments while exploring an unknown environment. Polar rovers have provided CReSIS the unique opportunity to safely and remotely acquire radar and seismic data. This data are being used to study the effects of climate change on the ice sheets and the inherent impact of sea level rise on our planet. The University of Kansas and CReSIS will continue advancing these systems for polar and planetary exploration and subsurface mapping.

**Acknowledgements** The authors would like to thank Dr. Georgios Tsoflias and Anthony Hoch of CReSIS for helpful discussions on seismic surveying; Hans Harmon for his work in development of the polar rovers; and the ANTS/TETwalker team associated with the NASA Goddard Space Flight Center for discussions about the TETwalker design. This material is based upon work supported by the National Science Foundation under Grant No. ANT-0424589. Any opinions, findings, and conclusions or recommendations expressed in this material are those of the authors and do not necessarily reflect the views of the National Science Foundation.

# Chapter 2
# Guidance and Control of Formation Flying Spacecraft

F. Y. Hadaegh, G. Singh, B. Açıkmeşe, D. P. Scharf and M. Mandić

**Abstract** A key element of NASA's future space exploration is high precision formation flying (FF) for space interferometry. Precision FF has never been attempted before and poses new and significant challenges to the underlying control system. While the guidance and control (G&C) methodologies of single spacecraft for traditional planetary flyby and orbiter missions are well-understood, the G&C of FF missions is fundamentally different. The FF systems require new control systems, architectures, and greater levels of autonomy to meet expected precision performance in the presence of environmental disturbances, plant uncertainties and more complex system interactions. This chapter will trace the motivation for these changes and will layout approaches taken to meet the new challenges.

Fred Y. Hadaegh
Jet Propulsion Laboratory, California Institute of Technology, Pasadena, CA 91109,
e-mail: Fred.Y.Hadaegh@jpl.nasa.gov

Gurkirpal Singh
Jet Propulsion Laboratory, California Institute of Technology, Pasadena, CA 91109,
e-mail: Gurkipal.Singh@jpl.nasa.gov

Behcet Açıkmeşe
Jet Propulsion Laboratory, California Institute of Technology, Pasadena, CA 91109,
e-mail: Behcet.Acikmese@jpl.nasa.gov

Daniel P. Scharf
Jet Propulsion Laboratory, California Institute of Technology, Pasadena, CA 91109,
e-mail: Daniel.P.Scharf@jpl.nasa.gov

Milan Mandić
Jet Propulsion Laboratory, California Institute of Technology, Pasadena, CA 91109,
e-mail: Milan.Mandic@jpl.nasa.gov

## 2.1 Introduction

Formation flying spacecraft refers to a set of spatially distributed spacecraft , capable of autonomously interacting and cooperating with one another (Figure 2.1). Many of NASA's future Earth and Space science missions involve formation flying. For Earth science applications, formation-flying spacecraft will enable a next generation of instruments for understanding the Earth and the effect of natural and human-induced changes on the global environment. For example, formations will enable distributed sensing for Earth gravity mapping, distributed atmospheric data collection, magneto-spheric studies, co-observations, and global communication systems. It will become possible to deploy large numbers of low-cost, Earth-orbiting minia-turized spacecraft/instruments with the opportunity for introducing new members to the formation for expansion, for upgrading technologies, or to replace a failed member. Similarly, several space science missions (e.g., Terrestrial Planet Finder, Interferometers (see Figure 2.2) and Terrestrial Planet Imager (103)) include dis-tributed instruments such as a large phased array of lightweight reflectors and an-tennas and long, variable baseline space interferometers. The group of collector and combiner/integrator spacecraft would form a variable-baseline optical space inter-ferometer for a variety of science applications from exo-Earth detection to imaging the event horizon of a black hole. The elements of the formation flying system must act collaboratively as a single unit, which represents a common system to perform a task. The quality of the collective behavior of all the spacecraft in the formation will determine the quality and the magnitude of the science return. Formation fly-ing spacecraft must conform to extremely stringent control and knowledge require-ments. Such precise requirements have never existed before. For example, the con-trol system for space interferometry, for example, must provide precision station-keeping from coarse requirements (relative position control of any two spacecraft to less than 1 cm, and relative attitude control of 1 arcmin over a large range of separation from a few meters to tens of kilometers) to fine requirements (nanometer relative position control, and .01 milliarcsec relative attitude control). Conformance to such precise performance metrics presents new challenges, not only in the areas of guidance, estimation, and control, but also in the areas of dynamic modeling of the formation flying spacecraft and their environment. It will be crucial to better

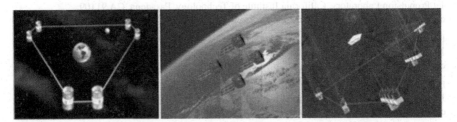

**Fig. 2.1** Earth Orbiting and Deep Space Formation Flying Spacecraft Interacting with One An-other for Co-observations

understand and model physical effects that would have been deemed unimportant or secondary for less precise spacecraft control applications.

The development of techniques to ensure stability, performance, and efficiency throughout various stages of formation flying mission has been an active area of research in recent years (161), (135) (163), (7), (160), (150). These researches have focused on formation modeling, attitude coordination, formation geometry, autonomous formation reconfiguration, time constraints, fuel efficiency, maneuver optimality and collision avoidance (151), (152). The purpose of this chapter is to provide an overview of the fundamental issues in areas of modeling, estimation, guidance and control of formation flying spacecraft with references to recent theoretical and experimental research developments in this area.

## 2.2 Modeling and Simulation

The conception of models and the design of simulation techniques for formation flying spacecraft poses significant challenges compared to those of conventional spacecraft. An efficient description is needed of the absolute and relative translational and rotational dynamics of the entire formation. The formation geometry can be expressed in terms of the states of a spacecraft in the formation and the states of the remaining spacecraft relative to the designated reference spacecraft. This naturally introduces an effective coupling between all spacecraft states that must be maintained throughout the formation. Different scales of motion occur simultaneously in a formation: translations and rotations of the formation as a whole (macro-

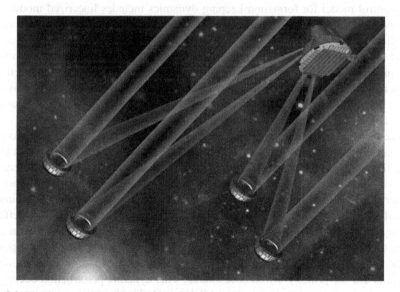

**Fig. 2.2** The Terrestrial Planet Finder Formation

dynamics), relative rotation and translation of one formation member with respect to another (relative dynamics), and formation member flexibility (micro-dynamics). A challenge is to incorporate these modes of motion into a usable reduced order model. The derivation of reduced order models for control, and the need to conveniently represent external perturbations and modeling uncertainties entering the model of a formation, are also unsolved problems.

From a dynamical standpoint, a formation of spacecraft is characterized by a wide dynamic range (from less than 1 Hz in the spacecraft dynamics to KHz in the operation of the instrument synthesized by the formation), and by spatial scales ranging from nanometers to thousands of kilometers. Techniques to model such systems do not yet exist. Significantly higher fidelity modeling and computational architectures will be needed to develop, test, and validate distributed spacecraft missions. Never before has the modeling fidelity been more strongly driven by the precision formation flying requirements. The formation can be thought of a virtual truss in which the stiffness and dissipation levels of the connecting links are dictated by the relative sensing and actuation between two or more neighboring spacecraft. The dynamic model of this virtual truss suffers from undesired deformation modes caused by sensor noise, actuator non-linearity, dynamic uncertainties, and environmental disturbances. Some of these perturbations are stochastic in nature, others are well predicted by deterministic models. In light of the unprecedented, extremely fine performance requirements, a comprehensive modeling of all uncertainties becomes more important for formation flying spacecraft than for conventional spacecraft. Specifically, in a low Earth orbit, orbital dynamics and environmental disturbances introduce additional strong, non-uniform, nonlinear dynamic perturbations to each spacecraft in the formation. Whereas the full nonlinear model is used to validate performance requirements, the formation control model is used to develop control laws. The control model for formation-keeping dynamics includes linearized models of the open-loop dynamics of each spacecraft, controller- and estimator-induced state coupling, sensor/actuator dynamics, sensor/actuator location and locations and nonlinear models of the orbital dynamic effects of Earth and Sun. The dynamic model also incorporates nonlinear models of the environmental perturbations induced by Earth magnetic field, Earth radiation pressure, solar pressure, harmonics of gravitational potential, gravity gradient disturbances, thermal effects of solar illumination and Earth albedo.

The formation geometry can be severely impacted by the environmental effects in low Earth orbits. Atmospheric drag is predominant over all other effects up to an altitude of about 500 kilometers. For large formations in low Earth orbit, the spacecraft at lower altitudes will experience stronger retarding forces as compared to those at higher altitudes. Furthermore, the formation geometry is influenced by predominant perturbations induced by high Knudsen number flow at higher altitudes. The effect of the atmospheric drag can be classified as a variable spatial perturbation on the formation that can only be accounted for by realizing appropriate state coupling of the formation geometry. After drag, higher-order harmonics of the gravitational potential are the primary disturbance source. This dynamic perturbation occurs at a frequency commensurate with the orbital period. The formation states are then affected directly by different disturbance frequencies corresponding to the various harmonics of the gravitational potential. A representative dynamic model of this phenomenon must be included in the formation control model to explore the impact on the formation geometry.

The geomagnetic field has a non-negligible intensity in low Earth orbits. The field behaves as a disturbance source to the formation and introduces non-uniform perturbations to the formation geometry. This is particularly important if each spacecraft in the formation is carrying magnetic loops (97) or is an element of a conductive loop with the ionospheric plasma (96), (89). The formation, in this case, acts as an electrical conductor, or equivalently, behaves as a cluster of dipoles that can be polarized in response to variations of the external electromagnetic field. As a result, the variability of the external field that varies proportionally with the size of the formation introduces variable differential forces and torques acting on different states of the formation.

Solar pressure is the dominant external perturbation on deep-space formation flying spacecraft. The solar pressure effect is analogous to Earths albedo, which acts on each spacecraft like a radiation pressure term at low altitude. Accurate modeling of obscuration effects, implying a detailed shadowing analysis, become essential.

In addition to direct solar pressure, radiative input on each spacecraft from different sources such as Earths albedo and emitted radiation of nearby spacecraft, needs to be taken into account for validation of tight formation requirements. Although these effects may be negligible for micro- or nano-spacecraft, larger-size spacecraft are particularly sensitive to it in proportion to the exposed area and relative distance. Impingement of thruster plumes from neighboring spacecraft must be included (196).

Turning from environmental modeling to spacecraft subsystem modeling, information processing on formation flying spacecraft is inherently distributed by nature. A formation model must include the latencies and bandwidth limitations associated with inter-spacecraft communications (162). Single spacecraft applications are immune to such considerations and limitations. Simulation of distributed spacecraft must also address a large range of spatial and temporal scales. It is in effect a multiple-scale problem, a solution to which will require a new class of numerical algorithms with special demands on accuracy and stability.

As with any spacecraft application, closed-loop, real-time testing of the flight system is critical for reducing mission risks and costs. Satisfaction of the real-time performance requirements in the presence of significantly higher modeling complexity and increased processing needs of the formation flying control system poses a challenge in the areas of development of distributed computing architectures, novel formation modeling methodologies, and efficient, higher-order, numerical integration algorithms.

## 2.3 Guidance and Control Architectures

Traditional single spacecraft applications have limited degrees of freedom in which a guidance and control architecture may be implemented. In particular, there is generally a single spacecraft with a single processor that must carry out the processing needed for all guidance and control functions. Co-operating spacecraft have physically separated platforms and processors, which brings about the possibility of shared, but distributed responsibilities. Many architecture implementation combinations are possible in a formation flying network, and the number of possible implementations increases with the number of spacecraft in the formation. It is clear that certain guidance and control architectures will be more robust, more computationally efficient and more cost-effective than others. The problem therefore is to find which architectures are most suitable. The chosen architecture will dictate, to a large extent, the sensor suite needed on each spacecraft. A centralized architecture is one where most of the information processing related to formation guidance, estimation, and control functions takes place on a single spacecraft. On one extreme, such an architecture would function by explicitly commanding all degrees of freedom of each spacecraft in the formation. Although it may make sense to do this for certain functions, it is not desirable to have such an arrangement for reasons of functional redundancy, robustness, and processing limitations. Additionally, communication can grow large, limiting controller performance in larger formations. A decentralized architecture, on the other hand, is one where formation functions, sensing, and information processing related to formation guidance, estimation, and control are the responsibility of physically separated platforms. Now, the stability of decentralized control algorithms must be considered. Although, it will be desirable for the standpoint of functional redundancy to adopt a decentralized architecture, centralized formation knowledge and planning is absolutely essential for the optimal planning of certain motions.

Figure 2.3 illustrates the exchange of information between two spacecraft in a Leader-Follower Architecture. In addition to the traditional spacecraft functions of inertial-relative attitude estimation, ephemeris propagation, control allocation and attitude commanding and control, several additional functions unique to formation-flying spacecraft are required. Autonomous collision avoidance is one such key capability. A centralized guidance architecture allows optimal, collision-free trajectories to be determined. Here, by "Centralized" we mean planning of relative translational degrees-of-freedom occurs on one, the Leader spacecraft. This is the responsibility of the "Translation Guidance" element in Figure 2.3 . Attitude planning ("Attitude Guidance" in Figure 2.3) on the other hand is performed more naturally in a local or a decentralized structure.

Formation state estimation is another formation-flying-specific function. Although not absolutely essential, it is highly desirable (functional redundancy, fault tolerance) to have local knowledge of relative states of all formation spacecraft, that is, an estimate of relative positions and velocities between all formation pairs in inertial space. The required relative state/measurement data is shared between formation elements to facilitate a formation state estimate. Formation control in this architec-

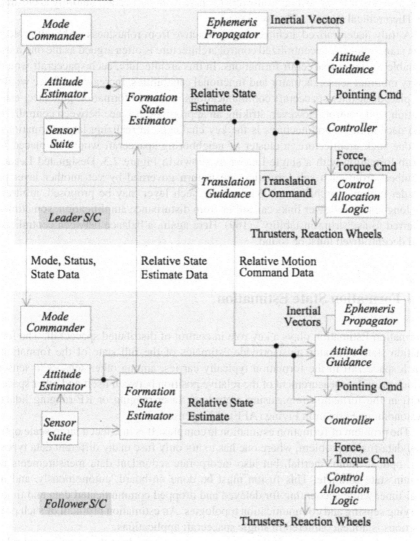

• Formation Command

Fig. 2.3  A Leader-Follower Guidance, Navigation, and Control Architecture for Formation-Flying

ture is rendered as a path-following exercise in each of the six local degrees-of-freedom. The Control Allocation Function is a more traditional spacecraft function which ensures that the forces and torques required to track some desired path are realized by a suitable combination of actuation elements (thrusters, reaction wheels). Note that, for thruster-only actuation, 6 DOF thrust allocation algorithms are needed to avoid over-conservatism. The pair-wise linking depicted in Figure 2.3 exists between all Leader-Follower pairs and there may be several such pairs in a formation. It is also possible for several Followers to share a single Leader. Stability is ensured by Hierarchical Systems Theory (184).

A truly decentralized architecture is attractive from robustness and fault protection standpoints. A decentralized control architecture is often argued as the one most suitable for large spacecraft formations. In this architecture, each spacecraft would carry minimal sensor, actuator and functional capabilities. Increased reliance would be placed on inter-spacecraft communications to enable formation guidance, estimation, and control. However, striking an appropriate balance between centralized and decentralized architectures is the key challenge in realizing large formations. In one such architecture, a cluster of neighboring spacecraft would be placed in a sub-formation with a single Leader as shown in Figure 2.3. Designated Leader members from each sub-formation are in-turn governed by yet another layer of Leader-Follower architecture. While many such layer may be proposed, arbitrarily long leader-follower links can suffer from disturbance amplification, sometimes referred to as "string instability" (169). Here again, a balance between centralized and decentralized must be found.

## 2.4 Formation State Estimation

Formation Estimation plays a key role in control of distributed spacecraft. The formation state estimator must provide estimates of the full state of the formation. Each spacecraft in the formation typically carries, among other sensors, a sensor which provides measurement of the relative position between itself and other spacecraft in the formation (e.g., using metrology, laser-ranging or RF-ranging lidars, Autonomous Formation Flying (AFF) (177) sensor).

The problem of formation estimation is complex. It is in effect a large-scale optimal data-fusion problem, where one has to not only fuse many different data types, e.g. optical, radio, inertial, but also incorporate redundant data measurements to obtain state estimate. This fusion must be done on-board, autonomously, and in real-time, while accounting for delayed and dropped communicated data and time-varying sensing and communication topologies. An estimation problem of such proportions is not encountered in single spacecraft applications.

To fully appreciate the challenges involved, we introduce the translational relative state estimation problem for formations on a circular orbit. The formation relative translational state is defined as the vector of positions and velocities of each spacecraft relative to a reference spacecraft in the formation, which is designated with the index 1 without loss of generality. Specifically, (see (5) for a derivation)

$$x = \begin{bmatrix} p_{12,1} & \cdots & p_{1N,1} \,|\, p_{12,2} & \cdots & p_{1N,2} \,|\, \cdots \,|\, v_{12,3} & \cdots & v_{1N,3} \end{bmatrix}^T \qquad (2.1)$$

where $N$ is the number of spacecraft in the formation, and

$$p_{ij} = \begin{bmatrix} p_{ij,1} \\ p_{ij,2} \\ p_{ij,3} \end{bmatrix}, \quad v_{ij} = \begin{bmatrix} v_{ij,1} \\ v_{ij,2} \\ v_{ij,3} \end{bmatrix}, \quad i = j,\ i = 1,...,N \quad 1,\ j = 2,...,N,$$

where $p_{ij,k}$ ($v_{ij,k}$), $k = 1,2,3$, is the $k^{th}$ coordinate of the position (velocity) of the $j^{th}$ spacecraft relative to the $i^{th}$ spacecraft. The discrete time dynamics for the relative formation state can be given by

$$x_{k+1} = Ax_k + B(u_k + w_k) \tag{2.2}$$

where $k = 0,1,2,....$ are the time indices,

$$A = e^{A_0 \Delta t} \quad I_N \ 1, \qquad B = \int_0^{\Delta t} e^{A_0(t\ \ \tau)} B_0 d\tau \quad I_N \ 1$$

$w_k$ is assumed to be zero mean white noise process with cov $w_k = Q_k$,

$$A_0 = \begin{bmatrix} 0_3 & I_3 \\ \omega^2 D_0 & \omega S_0 \end{bmatrix}, \quad B_0 = \begin{bmatrix} 0_3 \\ I_3 \end{bmatrix}, \quad D_0 = \begin{bmatrix} 3 & 0 & 0 \\ 0 & 0 & 0 \\ 0 & 0 & 1 \end{bmatrix}, \quad S_0 = \begin{bmatrix} 0 & 2 & 0 \\ 2 & 0 & 0 \\ 0 & 0 & 0 \end{bmatrix},$$

$\omega = \sqrt{\mu/R^3}$, $R$ is the radius of the orbit and $\mu$ is the gravitational parameter of the primary body, $I_m$ and $0_m$ are the $m$ $m$ identity and zero matrices, and    is the Kronecker product. Furthermore, $u$ and $w$ are the control and disturbance inputs relative to the spacecraft 1. That is,

$$u = \begin{bmatrix} u_{12,1} & \cdots & u_{1N,1} & | & \cdots & | & u_{12,3} & \cdots & u_{1N,3} \end{bmatrix}^T$$

where $u_{ij} = u_j$ $u_i$ with $u_i$ is the control input of the $i$th spacecraft and

$$u_{ij} = \begin{bmatrix} u_{ij,1} & u_{ij,2} & u_{ij,3} \end{bmatrix}^T$$

The same notation applies for the process noise vector $w$.

Measurements for estimation can be obtained by a variety of formation sensors and can be of different types. Without loss of generality, we assume that the measurements from each sensor gives a relative position vector between a pair of spacecraft in the formation (range-only measurements are treated subsequently). At any discrete time instance $k$, let $E_k$ be a matrix describing the $m_k$ available relative position measurements as follows: Each row of $E_k$ corresponds to an independently obtained relative position measurement $p_{ij}$ with $j$th entry of the row $+1$, $i$th entry 1, and the rest of the entries being zeros.

Throughout this section, we assume that the measurement links that exist at any given time index form a *connected sensing graph* (see (58) for a definition). Intuitively, this assumption means there are enough measurements to construct an "effective measurement" between any two spacecraft via vector addition of measurements. Given this assumption, one can show that (5), the vector of all measurements can be given by excluding measurement noise,

$$y_k = \underbrace{\left[ I_3 \quad E_k T^T (TT^T)^{-1} \quad 0_{3(N-1)} \right]}_{:= C_k} x_k \qquad (2.3)$$

where $C_k \in \mathbb{R}^{3m_k \times 6(N-1)}$, and $T \in \mathbb{R}^{(N-1) \times N}$ matrix given by

$$T = \begin{bmatrix} 1 & 1 & 0 & \dots & 0 \\ 1 & 0 & 1 & 0 & \vdots \\ \vdots & \vdots & \ddots & \ddots & 0 \\ 1 & 0 & \dots & 0 & 1 \end{bmatrix}.$$

This matrix encodes the relative state definition with respect to spacecraft 1. Other definitions are possible, resulting in different T matrices. Once we have the measurement model given by the equation (2.3), the following system model is formed to be used in the formation estimator synthesis

$$\begin{aligned} x_{k+1} &= Ax_k + B(u_k + w_k) \\ y_k &= C_k x_k + n_k \end{aligned} \qquad (2.4)$$

where $w_k$ and $v_k$ are independent zero mean white noise processes with cov $w_k = Q_k$ and cov $n_k = R_k$. This model of the system has linear-time-invariant state dynamics and the measured output is a linear but time-varying function of the state. $C_k$ is time-varying to capture time-varying sensor graphs, that is, which spacecraft are measuring which other spacecraft. One interesting fact is that $L_k := E_k^T E_k$ is the Laplacian matrix corresponding to the sensing graph at time index $k$ and $E_k^T$ is also known as the incidence matrix in graph theory.

When the sensing graph is connected the pair $(C_k, A)$ is observable (5), (147). Hence one can use a Kalman filter with this model and obtain the optimal estimates of the relative formation state. However, this process requires matrix inversions and multiplications that can be computationally demanding with large state and measurement vectors. Additionally, Kalman filtering is optimal when the measurement and process noise properties are known. The assumption may not be valid when there is inaccurate thruster activity or corrupted initial conditions, and in that case, the Kalman filter may converge slowly.

$\lambda$-Estimators (5) have been developed primarily to guarantee fast convergence to an ultimate error covariance where the decay rate is robust to the process noise uncertainties. Furthermore, since the estimator gains are computed off-line for each possible sensor graph and stored onboard, the computational complexity is substantially lower than that of Kalman filters. Before the $\lambda$-estimator is introduced, we present an assumption that the measurement matrix $C_k$ is in a finite set of possible matrices $H_1, ..., H_q$ for all $k$, that is, there is some integer valued function $\tau : \mathcal{Z}_+ \to 1, .., q$ such that

$$C_k = H_{\tau(k)}, \qquad k = 0, 1, 2, ...$$

Note that we do not know exact form of the function $\tau(k)$. In short, we know a priori possible sensor graphs but not which graph is in effect at each time $k$. However, during operation in real-time, $\tau(k)$ can be determined from the available measurements. Given this characterization of the measurement matrix, a $\lambda$-estimator has the following form with a set of constant gains $L_k$, $k = 1,...,q$, for each sensing topology,

$$\hat{x}_{k+1} = A\hat{x}_k + L_{\tau(k)}(H_{\tau(k)}\hat{x}_k \quad y_k) + Bu_k. \tag{2.5}$$

An issue in $\lambda$-estimators is the enumeration of all possible sensing topologies, that is, all possible $C_k$'s. One practical solution of this problem is choosing a number of nominal topologies that must exist during planned mission operations and designing the gain matrices for these topologies. Assuming that the sensing graph is connected for all times, any other measurement set resulting from an unaccounted topology can be compressed to a nominal topology that it contains via weighted least squares (see Figure 2.4). This approach assumes there is at least one minimally connected nominal topology (i.e., with a sensing graph that is a *tree*), so that the measurement vector corresponding to any other topology can be expressed as a linear function of the measurement vectors of this nominal topology.

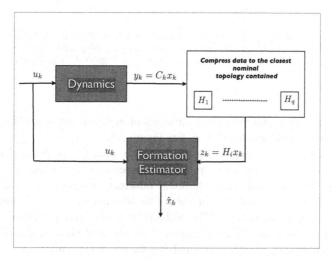

**Fig. 2.4** Data Compression to a Nominal Sensing Topology

Within this estimation framework, the formation state can be estimated by each spacecraft and, therefore the measurements and control actions must be exchanged between all spacecraft. Information exchange can introduce time delays due to the bandwidth limitations of the communication channels. However, as long as each spacecraft can communicate at least with one other spacecraft, there is a bound on the delay for any information to reach every spacecraft. The spacecraft can carry a formation estimator that produces estimates of the full delayed formation state with

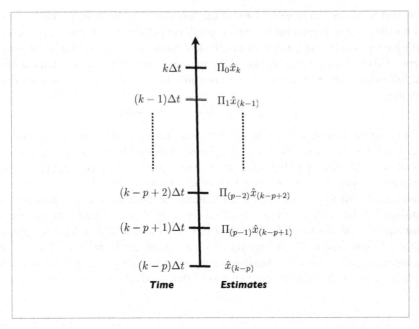

**Fig. 2.5** Estimation strategy with communication delays. $p\Delta t$ is largest delay due to communication network. $\Pi_i$, $i = 0, ..., p$ 1, are projection operators describing the locally observable state at a spacecraft, which satisfy ran$\Pi_0$ ran$\Pi_1, ..., $ran$\Pi_{(p\ 1)}$ $\mathbb{R}^{6(N\ 1)}$, where ran represents the range space of a matrix.

the worst case delay, and the partial estimates of more current states based on the instantaneously available information. See Figure 2.5.

We now consider a common measurement type, namely, one dimensional range measurements from an inter-spacecraft communication system or laser-range finder. While only range is measured, additional information is known based on sensor/communication antenna field-of-view. This information may be thought of as very coarse bearing information. The problem of determining positions relative to a spacecraft by using the range measurements and field-of-view information can be posed as: Compute $p_{1i}$, $i = 2, ..., N$ satisfying the following range relationships

$$
\begin{aligned}
p_{1i} &= d_{1i}, & i &= 2, ..., N, \\
p_{1j} \quad p_{1i} &= d_{ij}, & i &= j, i = 2, ..., N \quad 1, j = 2, ..., N,
\end{aligned}
\tag{2.6}
$$

where $d_{ij}$, $i = j$, $i = 1, ..., N$ 1 and $j = 2, ..., N$ are the all the inter-spacecraft distances measured (all the possible relative distances are measured), and the line-of-sight constraints given by

$$
n_{ij}^T p_{ij} \quad d_{ij} \sigma_{ij}, \qquad i = j, i = 1, ..., N \quad 1, j = 2, ..., N
\tag{2.7}
$$

where $n_{ij}$ is the known (from the spacecraft attitude measurements) unit vector describing the sensor bore-sight measuring the distance $d_{ij}$ and $\sigma_{ij}$ is determined by the field-of-view of the sensor. The system of equations and inequalities determined by (2.6) and (2.7) must be solved simultaneously to obtain the relative position vectors. It can be shown additional sensing, such as relative bearing angles between any pair of spacecraft, or spacecraft attitude maneuvers are needed for unique determination of these vectors.

## 2.5 Guidance and Control

By guidance we mean the planning of the attitude and translation of each spacecraft. Attitude is typically inertial-relative while translations are typically relative positions in inertial frame. The unprecedented complexity introduced by the required cooperation between several spacecraft flying in close proximity of one another poses a major challenge, especially in the path-planning area. Several new maneuver types, which are not needed by single spacecraft applications, must be addressed by formation flying algorithms. Some of these strongly couple attitude and translation path planning. While the coupling is completely absent in single spacecraft applications, for multi-spacecraft, guidance problem presents a new class of problems.

A formation-synthesized instrument is generally functional only when it has a specific configuration, that is, when each spacecraft in the formation is located at a prescribed position and attitude in relation to the other spacecraft. Such a configuration may be dynamic or static. To perform desired mission objectives, formations must often reconfigure, which is the process of reorienting, reshaping, and repositioning. This process involves each spacecraft in the formation translating to a new position and rotating to a new attitude. The attitude and translation of each spacecraft has to be planned between the current and the desired terminal values such that certain motion constraints are not violated. These motion constraints take the form of pointing, collision-avoidance, resource, actuation, and optimality constraints. The fundamental challenge in translational formation guidance is the non-convex collision avoidance constraint.

The general formation-flying path-planning problem for an N-spacecraft formation, with Linear Time-Invariant (LTI) relative dynamics is posed in the next section. This formulation encompasses deep space configurations and those in circular orbits.

### 2.5.1 Formulation of Optimal Path Planning Problem

The spacecraft are modeled as point masses in deep space or a circular Low Earth Orbit (LEO). We assume the following linearized dynamics for each spacecraft,

$$\dot{x}_j = A_c x_j + B_c u_i, \qquad j = 1, \ldots, N \qquad (2.8)$$

where $N$ is the number of spacecraft, $x_j \in \mathbb{R}^6$ is the state vector and $u_j \in \mathbb{R}^3$ is the control acceleration vector of the $i$th spacecraft.

The state vector is composed of position and velocity relative to the reference orbit in LEO and to an inertially fixed point in deep space:

$$x_j = \begin{bmatrix} r_j \\ v_j \end{bmatrix},$$

where $r_j \in \mathbb{R}^3$ is the position vector and $v_j \in \mathbb{R}^3$ is the velocity vector. The objective of a reconfiguration maneuver is to bring the formation to a desired configuration at time $t = T > 0$ from an existing configuration at time $t = 0$, which implies the following state constraints

$$
\begin{aligned}
x_1(0) &= x_{1,0} \\
x_{1j}(0) &= x_{1j,I}, \qquad j = 2, \ldots, N \qquad (2.9) \\
x_{1j}(T) &= x_{1j,F}, \qquad j = 2, \ldots, N
\end{aligned}
$$

where, for $j = 2, \ldots, N$, $x_{1j,I}$ and $x_{1j,F}$ are the initial and final states of all space-craft relative to the first spacecraft, $x_{1,0}$ is the initial state of the first spacecraft. By definition,

$$x_{ij} = x_j - x_i, \qquad j > i, \ i = 1, \ldots, N-1.$$

Note that the initial and final state are assumed to satisfy all other state constraints that will be listed from this point on.

There can be an additional constraint of one of the following forms for the first spacecraft's final position

$$
\begin{aligned}
x_1(T) &= x_{1,F} \qquad \text{or} \\
x_1(T) &\in \text{Co}\{a_1, \ldots, a_{m_1}\}
\end{aligned}
\qquad (2.10)
$$

where $m_1 > 1$ is a positive integer, $\text{Co}\{a_1, \ldots, a_{m_1}\}$ indicates the convex hull of the vectors $a_1, \ldots, a_{m_1}$. The equality in (2.10) constrains the final state of the first spacecraft to a prescribed value and the inequality in (2.10) bounds it to a region. In some cases, the absolute position of a formation is immaterial to the functioning of a formation-synthesized instrument. That is, $x_1(T)$ is free. In this case we do not impose final state constraint.

*Remark 2.1.* A set of the following form

$$X = \{x : x \in \text{Co}\{a_1, \ldots, a_m\}\},$$

where $a_1, \ldots, a_m$ are given vectors, can equivalently be expressed by a finite number of linear inequalities. We use the convex hull notation in some of the constraint descriptions for its notational compactness.

Relative state constraints are imposed in the following general form

$$Fx_{ij}(t) \quad Co \quad b_1, \ldots, b_m \quad , j > i, \quad i = 1, \ldots, N \quad 1, \quad t \quad (0, T). \tag{2.11}$$

For example, by choosing $F = C_v$ where

$$C_v = \begin{bmatrix} 0 & I \end{bmatrix}, \tag{2.12}$$

we can bound the relative velocity between each pair of spacecraft with (2.11).

The only control constraint considered here is a bound on available control acceleration

$$u_j(t) \quad U_j, \quad j = 1, \ldots, N, \quad t \quad [0, T]. \tag{2.13}$$

The last constraint is the collision avoidance constraint that makes the problem non-convex and NP-complete (40),

$$C_p x_{ij}(t) \quad R_{ij} > 0, \quad j > i, \quad i = 1, \ldots, N \quad 1, \quad t \quad [0, T], \tag{2.14}$$

where $R_{ij}$ is the minimum allowable distance between $i$th and $j$th spacecraft,

$$C_p = \begin{bmatrix} I & 0 \end{bmatrix}, \tag{2.15}$$

and $I$ and $0$ are the identity and zero matrices, respectively, of appropriate dimensions.

The formation reconfiguration trajectory planning problem may now be stated as follows:

$$\min_{u_j(\ ),\ j=1,\ldots,N} \phi^l \quad _\lambda \quad \text{subject to} \quad \{ (2.8), (2.9), (2.10), (2.11), (2.13), (2.14) \} \tag{2.16}$$

where $\phi^l \quad \mathbb{R}^N$ is defined by

$$\phi_j^l = \begin{cases} \int_0^T u_j(t) \; dt, & l = 1; \\ \int_0^T u_j(t) \; ^2 dt, & l = 2. \end{cases} \quad j = 1, \ldots, N, \tag{2.17}$$

and $\phi \quad _\lambda$ defines the norm of the vector $\phi \quad \mathbb{R}^N$ as follows

$$\phi \quad _\lambda = \begin{cases} \sum_{j=1}^{N} \phi_j, & \lambda = 1; \\ \sum_{j=1}^{N} \phi_j^2, & \lambda = 2; \\ \max_{j=1,\ldots,N} \phi_j, & \lambda = \infty. \end{cases} \tag{2.18}$$

Note that the cost in the equation 2.16 is also convex. The following table provides a physical insight into the cost function.

| $l \setminus \lambda$ | 1 | 2 | $\infty$ |
|---|---|---|---|
| 1 | Total fuel | - | Maximum fuel in a spacecraft |
| 2 | - | Total energy | Maximum energy in a spacecraft |

The relative motion planning problem, as posed above, is an extremely difficult multi-dimensional, multi-objective optimization problem whose solutions must be obtained numerically. Closed form solutions to the problem as posed here do not exist. Further, the dimensionality of the problem increases with the number of spacecraft in the formation. The need for autonomy and near real-time performance of planning algorithms presents significant challenges in the areas of numerical optimization. In the interest of keeping the problem manageable, certain motion constraints that may be considered "secondary" have been ignored in the problem statement. Anti-pluming, stray-light and thermal radiation constraints are also relevant but perhaps not as important as the ones mentioned in equations (2.9-2.11,2.13,2.14). Critically, by ignoring these extra constraints, translational and rotational path planning degrees-of-freedom are decoupled. Disregarding rotational degrees-of-freedom results in a considerable simplification.

Although the algorithm must execute on-board, the optimal path-planning problem does not require a solution in real-time, for it is possible to execute the proposed algorithm slightly ahead of time, in anticipation of the impending reconfiguration maneuver. Example inter-spacecraft separations for a five-spacecraft formation reconfiguration are shown below. In the five-spacecraft formation reconfiguration shown in the Figure 2.6, there are ten distinct spacecraft pairs. Correspondingly, the optimal reconfiguration must satisfy ten collision avoidance constraints. None of the inter-spacecraft separations must fall below 20 meters in this example. The dashed time histories in the plot on the left depicts these separations if trivial, straight line paths are traversed between the prescribed boundary conditions. Two such paths violate the 20 meter threshold. Optimal separations are depicted in blue, all of which observe the 20 meter constraint. The plot on the right depicts the time histories of cost functions associated with linear and optimal paths, respectively. Additional examples may be found in (159).

Another solution to the path-planning problem is proposed in (6). The problem solution is applicable to the deep-space as well as earth orbiting formations. The optimal control problem is discretized in time to obtain a finite dimensional parameter optimization problem. The collision avoidance constraints are imposed via separating planes between each spacecraft pair. A heuristic is introduced to choose these separating planes that leads to a convexification of the collision avoidance constraints. Convexification permits claims regarding deterministic convergence of the proposed algorithm. The resulting finite dimensional optimization problem is a second order cone program, for which a number of traditional algorithms can be used to compute the global optimum with deterministic termination and with a prescribed level of accuracy of convergence. This solution property makes it eminently suitable for a real-time implementation. In a more complex seven spacecraft example discussed in (6), twenty-one collision avoidance constraints are observed. In this example all inter-spacecraft separations are required to remain above 4 meters. The resulting separation time histories are depicted in Figure 2.7.

Reconfigurations with weighted-fuel expenditures across various spacecraft are also important in formation flying applications. Mission longevity dictates near-equal fuel expenditure across all formation-flying spacecraft. In the perfect case, all spacecraft will exhaust their propellant at the same time in the mission. A heterogeneous formation would typically require penalizing reconfiguration delta-V in inverse proportion to the spacecraft mass. Single spacecraft applications do not have to address this issue. The relative motion dynamics (2.8) are dictated not by absolute, but rather by differential accelerations. Therefore, it becomes possible to integrate appropriate fuel-equalization metrics in the cost function by also requiring the minimization of suitably-weighted acceleration differentials (21).

Flying multiple spacecraft in formation in a low Earth orbit poses additional challenges. First, a suitable reference orbit must be determined. Unforced orbits of multiple spacecraft in general do not remain in a stable formation. Perturbations due to atmospheric drag and oblateness have destabilizing effects. It is possible to obtain $J_2$-invariant orbits by sufficiently constraining the formation geometry.

Formation control can be realized by tracking reference guidance commands such as relative translations derived from the solution to the optimal formation re-

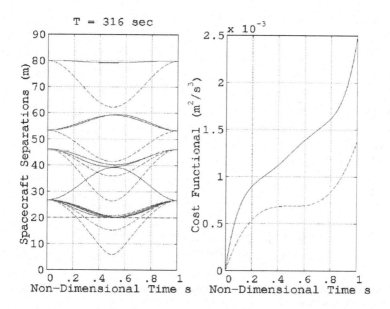

**Fig. 2.6** Inter-Spacecraft Separations for a Collision-Free Optimal Reconfiguration Maneuver for a Five-Spacecraft Formation. Performance without collision avoidance shown as dashed lines. Performance with collision avoidance shown as solid lines.

configuration problem or science based path such as virtual rigid body rotations
(143). This approach ensures optimality of control actions in the sense that the reference paths are optimal. Feedback control then provides essential disturbance rejection in the presence of environmental perturbations. It is envisioned that a single
controller will not be able to appropriately accommodate the large variations in spatial scales and control requirements of formation flying applications. Such problems
are traditionally handled by using different control laws for different performance
regimes. A problem central to any six degree-of-freedom spacecraft control application is that of control allocation so that appropriate six degree-of-freedom motion
is realized without incurring large control errors in any degree of freedom. Collective thrust actions allow desirable six degree-of-freedom motions to be realized.
The problem is simplified when attitude degrees of freedom are controlled by momentum exchange devices. For practical applications, however, it is envisioned that
thrusters will provide the initial coarse attitude control as well with momentum exchange devices providing the required attitude control in the fine attitude control
regimes.

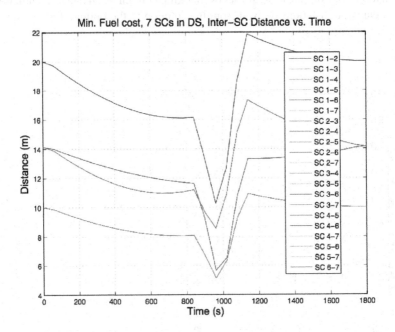

**Fig. 2.7** Inter-Spacecraft Separations for a Collision-Free Optimal Reconfiguration Maneuver for
a Five-Spacecraft Formation

## 2.6 Conclusions

Formation flying spacecraft enable new capabilities and new science. Space-borne optical interferometry is one such capability that has been identified as a critical technology for many of NASAs next generation science missions. The formation guidance, estimation and control problems in such applications are complex from the standpoint of higher-fidelity dynamic modeling of the spacecraft and its environment, computational complexities, GNC architectures, and information processing. In many cases, conventional spacecraft control approaches do not apply. The magnitude of these technical challenges, which increase with size of a formation, was discussed and some possible solutions to these problems were identified. However, significant advances in mathematical modeling, numerical simulation techniques, efficient decentralized control architectures, and real time solution to large-scale optimization problems in guidance and estimation will be needed to truly enable precision formation flying spacecraft.

## 2.7 Acknowledgement

This research was performed at the Jet Propulsion Laboratory, California Institute of Technology, under contract with the National Aeronautics and Space Administration.

# Chapter 3
# Adaptive Sampling for Field Reconstruction With Multiple Mobile Robots

Bin Zhang and Gaurav S. Sukhatme

**Abstract** When a scalar field, such as temperature, is to be estimated from sensor readings corrupted by noise, the estimation accuracy can be improved by judiciously controlling the locations where the sensor readings (samples) are taken. In this chapter, we formulate solve the following problem: given a set of static sensors and a group of mobile robots equipped with the same sensors, how to determine the data collecting paths for the mobile robots so that the reconstruction error of the scalar field is minimized. In our scheme, the static sensors are used to provide an initial estimate, and the mobile robots refine the estimate by taking additional samples at critical locations. Unfortunately, it is computationally expensive to search for the best set of paths that minimizes the field estimation errors (and hence the field reconstruction errors as well). We propose a heuristic to find 'good' paths for the robots. Our approach first partitions the sensing field into equal gain subareas and then we use a single robot planning algorithm to generate a path for each robot separately. The properties of this approach are studied in simulation. Our approach also implicitly solves a multi-robot coordination/task allocation problem, where the robots are homogeneous and the size of task set might be large.

## 3.1 Introduction

A sensor actuator network (also a robotic sensor network ), which consists of both static and mobile nodes, provides a new tool for measuring and monitoring the environment. On the one hand, with less energy consumed, the static sensor nodes

Bin Zhang
Department of Computer Science, University of Southern California, Los Angeles, CA 90089-0781, e-mail: binz@usc.edu

Gaurav S. Sukhatme
Department of Computer Science, University of Southern California, Los Angeles, CA 90089-0781, e-mail: gaurav@usc.edu

are able to provide high resolution temporal sampling. On the other hand, with the ability to move, the mobile nodes (henceforth robots) are able to change the spatial distribution of the sensor readings leading (using an appropriate algorithm) to a high density of readings in important areas. The key challenge is essentially an adaptive sampling problem - come up with trajectories that the robots can follow, sampling along which will improve the field reconstruction. In (203), we proposed an adaptive sampling algorithm for a system consisting of a set of static sensor nodes and one mobile robot, a robotic boat. The system, part of the NAMOS (168) project at USC (http://robotics.usc.edu/ namos), is used for measuring scalar fields, such as temperature, salinity and chlorophyll concentration. We have shown (203) that by combining optimal experimental design and path planning, we are able to achieve an improved estimation performance, i.e., a lower Integrated Mean Square Error (IMSE) with the same (finite) initial energy available to the mobile robot.

In (203), we assume that the scalar field to be reconstructed changes slowly. That is, during the time the mobile robot is sent out for a data collecting tour, the readings from the static sensors are still valid. However, this might not be true in practice since it takes a while for the mobile robot to finish a tour. One way to overcome this drawback is to use multiple mobile robots in parallel to accomplish the task. If we can generate 'good' paths for all the mobile robots and let them carry out the sampling task simultaneously, the speedup could be significant. Another advantage of a system with multiple mobile robots is energy efficiency. In many cases, the ideal distribution (leading to the best reconstruction of the field) of the sensor readings contains several clusters. Since normally, we already have static sensors covering the whole sensing field, the mobile robots may just need to take readings within each cluster. If only one mobile robot is deployed, it has to move between the clusters. If multiple mobile robots are used and the number of robots is more than the number of clusters, each robot only needs to stay within a cluster and the energy to move from one cluster to another could be saved.

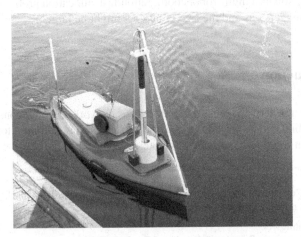

**Fig. 3.1** One of the robotic boats used in the NAMOS project at USC.

In this chapter, we investigate the problem of adaptive sampling using multiple mobile robots. Specifically, given a set of static sensor nodes deployed uniformly across the sensing field, and a team of mobile robots each with the same energy, how to exploit the information collected by the static sensors and coordinate the motion of the mobile robots so that error associated with the reconstruction of the underlying scalar field is minimized. Here we assume that all the mobile robots have the same energy consumption profile and the underlying scalar field is continuous and has finite second order derivative at any point.

As in the case of our prior work on the single robot problem (203), we use techniques from optimal experimental design to define a gain associated with each location and then apply a search algorithm to find 'good' paths for the mobile robots. The challenge in the multiple mobile robot case comes from the path planning. Even in the case of a single robot, to find the path with maximum sum of gains is NP-complete. This is called the Orienteering problem, which has been studied in the theoretical computer science and operating research communities. In the case of multiple mobile robots, the situation is even worse since the search space grows exponentially with the number of the mobile robots. Therefore, we will have to compromise performance by approximation so that the path planning can be done quickly enough to be realistically feasible for a real-world situation. In this chapter, we propose a divide and conquer approach to solve the problem. Once the gain of each location in the sensing field is computed, we partition the sensing field into subareas with equal gain. Now the path planning problem reduces to a set of problems. Each robot is assigned a subarea and a path planning algorithm for single robot within each region is applied. An advantage of this approach is that the path planning for a single mobile robots can be done in parallel since the subareas do not overlap.

This chapter is organized as follows. We first discuss related work in the next section. Then, the optimal experimental design is explained briefly in section 3.3. We discuss the partition strategy in section 3.4 and the simulations results are presented in section 3.5. We discuss future work and conclude in section 3.6.

## 3.2 Related Work

Adaptive sampling and actuated sensing have been studied in the sensor network community. Based on Wavelet theory, Willett (195) proposed an algorithm to extend the life time of a static sensor network by putting some sensors to sleep without significantly increasing the estimation errors. Rahimi (137) proposed a sequential algorithm for a single mobile node to minimize the estimation error with the limitation on the time or distance the mobile node travels. In his approach, more readings are taken in the place where there is higher residual. Krause (98) proposed a near-optimal algorithm to solve the problem of static sensor node deployment. However, it is assumed that the underlying phenomenon is a Gaussian Process and a dense deployment of sensors is needed to find the correlation between readings at any two

locations. Based on the same assumption, Singh (158) introduced an approximation algorithm to solve a problem that is similar to ours and also extended to the
case with multiple mobile robots by using sequential allocation. Our approach does
not assume previous dense deployment of the sensors, and is faster than sequential
allocation.

Another problem related to adaptive sampling with multiple mobile robots is the
multi-robot exploration and mapping. It is normally assumed that the robots are deployed in an unknown environment and no global information is available. In this
case, many techniques of coordinated exploration and mapping are based on the
idea of the frontier points or cells (65; 204; 157; 38). To assign frontier points to
individual robots, a market-based approach is used and the target locations are assigned through auctions (204; 157; 38). There are different ways to evaluate the
target locations for each robot. In the case where the energy consumption is important, the assignment with best tradeoff between energy and utility is chosen (38); Fox
took the uncertainty of the localization into consideration and proposed a strategy
that trades off between the frontier and the hypothesis. Normally, the coordination
strategies assign each robot one target location to visit. In the case multiple target
locations are to be assigned, a sequential allocation is used (204; 157). Stroupe (167)
proposed a value-based coordination algorithm, which trades off between dynamic
target observing, exploration, sampling and communications. However, the energy
consumption is not considered. In this chapter, we are looking at a slightly different
problem. First, we are going to use the static sensors to provide a coarse estimation
of the scalar field and hence rough global information is available. Second, multiple
target locations are to be assigned to each robot and each robot then visits these
locations sequentially. Finally, because of the existence of the static sensor nodes,
we assume that inter-robot communication is always possible using some sort of
multi-hop protocol and hence we do not consider communication as a constraint.

## 3.3 Adaptive Sampling

In this chapter, we assume that no parametric model is available for the scalar field
and non-parametric regression is appropriate. A Kernel estimator is one of the most
popular non-parametric regression techniques since it is easy to understand, analyze
and implement. Local linear regression is a kernel estimator where the value of the
scalar field at any location $\mathbf{x}$ is estimated by using weighted linear regression. It is
assumed that the closer two locations are, the higher the correlation between the
values. As a result, when the linear regression is applied to location $\mathbf{x}$, more weight
is given to the data points closer to $\mathbf{x}$ while less weight is given to the data points far
away. We assume a non-parametric model shown as equation 3.1.

$$y = m(\mathbf{x}) + \sigma(\mathbf{x})\varepsilon, \qquad (3.1)$$

where $\mathbf{x}$ is sensor location, $y$ is the corresponding sensor reading, $\sigma^2(\mathbf{x})$ is the variance of the noise and $\varepsilon$ is a random variable with zero mean and unit variance and are independent of $\mathbf{x}$. The local linear regression can be represented by equation 3.2.

$$\hat{m}(\mathbf{x},H) = e_1^T (X_x^T W_x X_x)^{-1} X_x^T W_x Y, \qquad (3.2)$$

where $X_x = \begin{bmatrix} 1 & (\mathbf{x}_1-\mathbf{x})^T \\ \vdots & \vdots \\ 1 & (\mathbf{x}_n-\mathbf{x})^T \end{bmatrix}$, $Y=[y_1,\cdots,y_n]^T$, $W_x = diag\{K_H(\mathbf{x}_1-\mathbf{x}), \cdots, K_H(\mathbf{x}_n-\mathbf{x})\}$ and $e_1 = [1,0,\cdots,0]$. The kernel $K_H(\mathbf{u})$ determines how much weight would be assigned to each data point and is normally defined by a base $K(\mathbf{u})$ and a matrix $H$,

$$K_H(\mathbf{u}) = |H|^{-1/2} K(H^{-1/2}(\mathbf{u})).$$

$H$ determines the size and shape of the neighborhood and $H^{-1/2}$ is called the bandwidth matrix.

If $K(\mathbf{u})$ satisfies $\int \mathbf{u}\mathbf{u}^T K(\mathbf{u})d\mathbf{u}$ is finite and $\int u_1^{l_1} \cdots u_d^{l_d} K(\mathbf{u})d\mathbf{u} = 0$ for all nonnegative integers $l_1,\cdots,l_d$ such that their sum is odd, it has been proved that the estimation error associated with the local linear regression is given by the following equation (146):

$$MSE\{\hat{m}(\mathbf{x};H)\} = \frac{C_1\sigma^2(\mathbf{x})}{n|H|^{1/2}f(\mathbf{x})}$$
$$+\frac{1}{4}C_2tr^2\{HH_m(\mathbf{x})\}$$
$$+o_p\{n^{-1}|H|^{-1/2}+tr^2(H)\}. \qquad (3.3)$$

where $n$ is the number of samples, $f(\mathbf{x})$ is the density function with $\int f(\mathbf{x})d\mathbf{x} = 1$, and $H_m(\mathbf{x})$ is the Hessian matrix of $m(\mathbf{x})$ and $C_1$ and $C_2$ are the constant depending on kernel $K(\mathbf{u})$.

If $n|H|^{1/2}$ is big enough and $H$ is small enough, the infinitesimal term is negligible. By applying the Lagrange-Euler differential equation, we can find the optimal bandwidth and corresponding IMSE as follows.

$$h = (\frac{dR(K)v(\mathbf{x})}{nf(\mathbf{x})\mu_2(K)^2 tr^2\{H_m(\mathbf{x})\}})^{\frac{1}{d+4}}, \qquad (3.4)$$

$$IMSE(\mathbf{X}_1,\dots,\mathbf{X}_{n0+n}) \propto \int (\frac{tr^d\{H_m(\mathbf{x})\}v^2(\mathbf{x})}{n^2\hat{f}^2(\mathbf{x})})^{\frac{2}{d+4}} d\mathbf{x}, \qquad (3.5)$$

where $\hat{f}(\mathbf{x}) = n^{-1}\sum_{i=1}^n K_H(\mathbf{X}_i-\mathbf{x})$ is the density function.

Assume that initially there are $n_0$ readings from static sensor nodes $(\mathbf{x}_1,y_1),\dots,(\mathbf{x}_{n_0},y_{n_0})$, the path of the mobile robot passes through the points $\mathbf{x}_{n_0+1}, \dots, \mathbf{x}_{n_0+n}$, then the optimal path should minimize IMSE. The IMSE can be estimated with the equation 3.5 and the corresponding density can be estimated by using $\hat{f}(\mathbf{x}) = n^{-1}\sum_{i=1}^n K_H(\mathbf{X}_i$

**x**) . Similar to the information gain defined in robot exploration literature, we define the gain for each point as follows.

$$G(\mathbf{x}) = IMSE(\mathbf{X}_1, \ldots, \mathbf{X}_{n_0}) \quad IMSE(\mathbf{X}_1, \ldots, \mathbf{X}_{n_0}, \mathbf{x}), \quad (3.6)$$

The gain associated with the path is defined as

$$G(p) = IMSE(\mathbf{X}_1, \ldots, \mathbf{X}_{n_0}) \quad (3.7)$$
$$IMSE(\mathbf{X}_1, \ldots, \mathbf{X}_{n_0}, \mathbf{X}_{n_0+1}, \ldots, \mathbf{X}_{n_0+n}).$$

Now, the problem is to find the path or paths collecting the most gains and hence a path planning problem needs to be solved.

## 3.4 Divide and Conquer

---
**Algorithm 1**: Adaptive Sampling With Multiple Mobile Robots
---
Construct the state graph $G=(V, E)$;
Collect readings from static sensors $r0$;
**for** *each vertex v    V* **do**
    Compute gain $g(v)$;
**end**
Partition $G$ into subgraphs $G_1, \ldots, G_m$;
**for** $i = 1$ *to m* **do**
    $\pi_i = ABFS(G_i, E)$;
    Collect new reading $r_i$;
**end**
Reconstruct the scalar field from $r_0$   $r_1$   $\ldots$   $r_m$;

---

### 3.4.1 Discretization

In the problem of adaptive sampling with static sensor nodes and multiple mobile robots, the main constraint is that the energy available to each mobile robot is limited. Therefore, path planning for the mobile robots needs to take into consideration the energy consumption model. The simplest model is to assume that the energy consumed by a mobile robot to move from location $A$ to $B$ is proportional to the length of the line segment connecting the $A$ and $B$. The energy consumption model used in this chapter is the one for the robotic boat of the NAMOS project (59) (shown in Figure 3.1). In this model, the state is the location and the orientation of the boat. The energy consumed for state transition not only depends on the distance between two states but also depends on the orientations of the two states. The details of the energy consumption model are described in (203) and will not be repeated here.

In our approach, we first construct a graph from the sensing field and the energy consumption model. Each vertex of the graph represents one state of the location of the mobile robot and the coordinates of any vertex are within the sensing field. The edges between vertices represent a state transition. The length of an edge represents the energy consumed for state transition. An important assumption is that all the mobile robots share the same energy consumption model and are equipped with the same sensors, i.e., the team of the robots is homogeneous. For each vertex of the graph, the Hessian matrix is estimated by using Local Polynomial Regression using the readings from the static sensors and the gain is computed by using equation 3.6. Note that the vertices with the same coordinates share the same gain. If the robot visits one location twice but in different orientations, it collects the gain once.

## 3.4.2 Graph Partition

Once the graph is constructed and the gain associated with each state is computed, we divide the graph into subgraphs to simplify the problem. The graph is partitioned in such a way that the sum of the gains of the vertices in each sub-graph are the same. The basic idea here is that with the same amount of energy consumed, the same amount of gain is achieved. Another constraint on the partition is that the boundary between two sub-graphs should be as straight as possible. Generally speaking, a complex boundary would result in more energy consumption. The complexity of the boundary can be measured by using the length of the boundary and the length of the boundary in turn can be measured by using the number of cuts of between sub-graphs. Therefore, the partition problem is to find a partition of $m$ sub-graphs so that the total of gain of all the nodes in each sub-graphs is the same, where $m$ is the number of mobile robots.

Graph partition is a well known NP-complete problem and there is no polynomial time algorithm to find the optimal partition. However, many approximation algorithms have been proposed. The approach used in this chapter exploits a multilevel paradigm (50). This approach consists of two stages. In the first stage, the graph is contracted until the size is less than a given threshold. Then, an interactive process of expansion and refinement is performed in such a way that the weight is balanced. This algorithm runs quickly with reasonable graph sizes.

### 3.4.3 Path Planning for a Single Robot

Once the graph is partitioned, the problem is reduced to a set of smaller problems each with a single mobile robot and many search algorithms can be used. The problem of finding the best path such that the gain collected along the path is maximized is called the Orienteering problem. This problem has been well-studied and many approximation algorithms are proposed. Most of the algorithms employ a prime and dual scheme and the typical one is proposed in (31; 42). In (203), A Breadth First Search (BFS) algorithm was proposed to find a approximate solution. Although the approximation factor is not good, the algorithm runs quickly and the performance is good. The approximate BFS is used here as the single mobile path planning algorithm. However, any other path planning algorithm can well be used here.

In summary, the algorithm is described in Algorithm 1, where $m$ is the number of mobile robots, $E$ is the initial energy for the mobile robots, $G$ is the state graph, $G_i$ is a subgraph and $\pi_i$ is the path generated in subgraph $G_i$.

## 3.5 Simulations

We carried out a series of simulations in a unit square. Three scalar fields are used in the simulations. They are shown in Figure 3.2 and their equation are below.

$$r = \frac{1}{1 + exp(\frac{1}{2}x^2 \quad y + \frac{1}{5})} \tag{3.8}$$

$$r = exp(\quad \frac{1}{2}(4x^2 \quad 10y + 3)) \tag{3.9}$$

$$r = exp(\quad \frac{(5x \quad 1)^2 + (5y \quad 1)^2}{2}) \tag{3.10}$$

In the first field, there is a boundary across the sensing field and it separates the high and low values. The second scalar field has a ridge across the sensing field. In our simulation, we smooth the readings by applying a local linear regression on the data to reduce the noise associated with it. As a result, in the place where the scalar field is not symmetric, the place with the maximum trace of the Hessian matrix would be moved. The biggest difference between field 1 and the other two is that in field 1, the estimation of the Hessian matrix would not only have errors in magnitude but also in location. Therefore, we expect higher estimation errors in reconstructing field 1.

When sensor readings are taken, either from the static sensor nodes or the mobile robots, we assume a Gaussian noise associated with the readings. Note that all the scalar fields vary from 0 to 1 and we use the ratio of the noise to the variation of the scalar field to describe the noise level. In the simulations, we use a coffeehouse design (117) to determine the locations of the static sensor nodes. The simulations are performed in groups. The sensing field is discretized into a graph consisting of 100

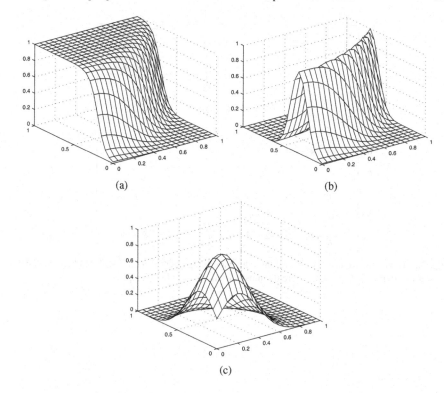

**Fig. 3.2** Scalar fields used in the simulation.

vertices uniformly distributed in the unit square. For each set of initial sensor readings, we estimate the Hessian matrix for each vertex in the graph and then compute the corresponding information gain. Then the graph is partitioned into subgraphs with equal gain. After that, the path planning procedure is called 25 times for each initial energy level and the new readings are collected. The scalar field is estimated and the IMSE is computed for 25 times. The whole process is in turn carried out over 10 trials.

One example of the paths planned for all the robots is shown in Figure 3.3. The underlying phenomenon is defined by equation 3.10. The paths are generated in such a way that more readings are taken in the lower left part of the sensing field, where the trace of the Hessian matrix is much higher than other places. Figure 3.4 shows how the IMSE changes with increased initial energy available to the mobile robots in three different settings. In all simulations, the IMSE decreases when the initial energy available to the mobile robots increases. We study the effect of two parameters, the number of static sensors and the noise level. The simulation results show that both of them have an effect on the performances. The simulation with the best performance is shown in Figure 4(a), where 50 static sensors are used and the noise level, 5%, is low. When the initial energy is 1.6 units, which means approxi-

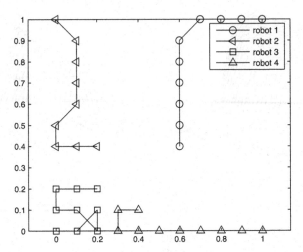

**Fig. 3.3** The paths planned for all 4 robots. The underlying phenomenon is scalar field 3. Note the cluster of samples in lower left where the most variation in the field happens.

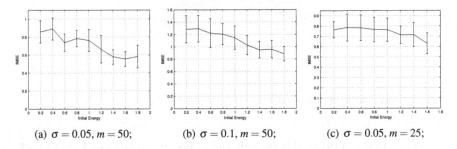

(a) $\sigma = 0.05, m = 50$;         (b) $\sigma = 0.1, m = 50$;         (c) $\sigma = 0.05, m = 25$;

**Fig. 3.4** The IMSE of the reconstruction of scalar field 1 with different noise levels ($\sigma$) and different number of static sensor nodes ($m$).

mately 36 new readings are taken, the IMSE decreases to 64% of the initial IMSE. When the noise level of the readings increases, not only does the absolute value of the IMSE increases, the rate at which the IMSE decreases also decreased. As shown in Figure 4(b), with the same initial energy of 1.6, the IMSE is still approximately 75% of the initial IMSE. Fewer initial static sensors also reduce the rate of IMSE decrease, as shown in Figure 4(c), where there only 25 static sensors to provide initial readings. The reason for this is that both parameters affect the accuracy of the estimation of the Hessian matrix. When the number of initial readings is small or the sensor noise is high, the error in the estimation of the Hessian matrix is high, which in turn causes more readings to be taken at the non-critical places.

Figure 3.5 shows the results of simulations on the other two scalar fields. Both sets of simulations are performed with the same noise level and the same number of static sensors as in Figure 4(a). As we discussed above, both of the scalar fields are

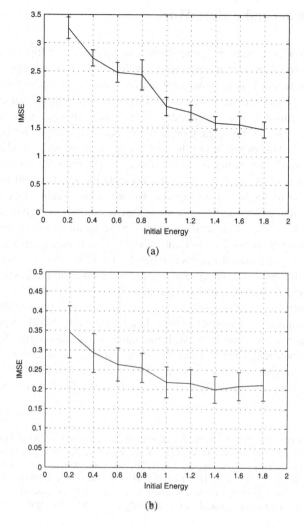

**Fig. 3.5** The change in IMSE of the reconstruction of scalar field 2(a) and 3(b) with the availability of increased initial energy to the robots.

symmetric and hence estimation of the Hessian matrix only has error in magnitude with much less error on the location. Therefore, IMSE in both situations has a high rate of decrease, which is obvious in Figure 3.5.

## 3.6 Conclusion and Future Work

In this chapter we presented a simple strategy to coordinate multiple mobile robots to take sensor readings so that errors associated with the reconstruction of a scalar field would be reduced. Local linear regression is used to estimate the scalar field and an optimal experimental design is used to define the information gain of each location. The sensing field is partitioned into subareas with equal gain. Within each subarea path planning for a single mobile robot is used to generate the path for each individual robot. This properties of the strategy are studied using simulation.

However, equal gain is not the only strategy to partition the sensing field. For example, another even simpler strategy, partitioning the sensing field into subareas with equal area might be a good option. We performed preliminary simulations on the latter strategy and the results show that it might be competitive with the equal gain strategy. We are currently working on the detailed analysis on the second strategy. Figure 3.6 shows the preliminary simulation results from the equal area strategy. Figure 6(a) shows the paths planned for the four robots in one simulation. Compared with Figure 3.3, fewer readings are to be taken in the lower left part of the sensing field. However, equal area strategy is still able to achieve a estimation error that is very close to the equal gain strategy, as shown in Figure 6(b).

We plan to apply this approach to the robotic boats in the NAMOS project. Currently, there are two robotic boats with the same configuration and they are used for measuring physical, chemical and biological parameters on the water surface as well as in depth. Our plan is to test our strategy on the robotic boats in a lake or harbor where there are reasonable variations in the scalar field, such as temperature, on the surface.

**Acknowledgements** The authors thank David Caron, Jnaneshwar Das, Amit Dhariwal, Mark Hansen, David Kempe, Carl Oberg, Arvind Pereira, and Beth Stauffer for their comments and help with the experiments. This work is supported in part by the National Science Foundation (NSF) under grants CNS-0325875, IIS-0133947, EIA-0121141 and grants CCR-0120778, ANI-00331481 (via subcontract). Any opinions, findings, and conclusions or recommendations expressed in this material are those of the authors and do not necessarily reflect the views of the National Science Foundation.

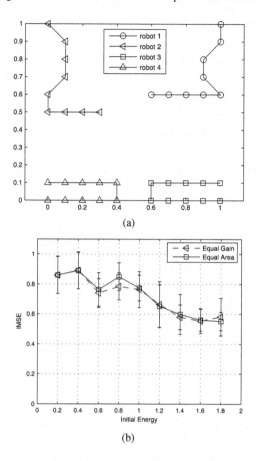

**Fig. 3.6** The preliminary simulation results of the equal area strategy.

# Chapter 4
# Grasping Affordances: Learning to Connect Vision to Hand Action

Charles de Granville, Di Wang, Joshua Southerland, Robert Platt, Jr. and Andrew H. Fagg

**Abstract** When presented with an object to be manipulated, a robot must identify the available forms of interaction with the object. How might an agent automatically acquire this mapping from visual description of the object to manipulation action? In this chapter, we describe two components of an algorithm that enable an agent to learn a grasping-oriented representation by observing an object being manipulated by a human teacher. The first component uses the sequence of image/object pose tuples to acquire a model of the object's appearance as a function of the viewing angle. We identify visual features that are robustly observable over a range of similar viewing angles, but that are also discriminative of the set of viewing angles. Given a novel image, the algorithm can then estimate the angle from which the object is being viewed. The second component of the algorithm clusters the sequence of observed hand postures into the functionally distinct ways that the object may be grasped. Experimental results demonstrate the feasibility of extracting a compact set of canonical grasps from this experience. Each of these canonical grasps can

Charles de Granville
Symbiotic Computing Laboratory, School of Computer Science, University of Oklahoma, Norman, OK 73019, e-mail: chazz184@gmail.com

Di Wang
Symbiotic Computing Laboratory, School of Computer Science, University of Oklahoma, Norman, OK 73019, e-mail: di@cs.ou.edu

Joshua Southerland
Symbiotic Computing Laboratory, School of Computer Science, University of Oklahoma, Norman, OK 73019, e-mail: Joshua.B.Southerland-1@ou.edu

Andrew H. Fagg
Symbiotic Computing Laboratory, School of Computer Science, University of Oklahoma, Norman, OK 73019, e-mail: fagg@cs.ou.edu

Robert Platt, Jr.,
Dexterous Robotics Laboratory, Johnson Space Center, NASA, Houston, TX, e-mail: robert.platt-1@nasa.gov

then be used to parameterize a reach controller that brings the robot hand into a specific spatial relationship with the object.

## 4.1 Introduction

Manipulating one's world in very flexible ways is a skill that is shared only by a small number of species. Humans are particularly skilled at applying their manipulation abilities in novel situations using a range of effectors, from hands and other parts of the body, to tools. How can robots come to organize and learn knowledge representations for solving grasping and manipulation problems in unstructured environments? J. J. Gibson (1966, 1977) suggests that these representations should be partitioned into *what* can be done with particular objects and *why* an object should be manipulated in a certain way. The first of these, which Gibson terms *object affordances*, captures the details of what can be done with the object by the agent. The latter captures information about how individual manipulation skills are to be put together in order to solve a specific task. The task-neutral affordance representation is important in that it can provide an agent with a menu of actions/activities that are possible with a given object – whether the current task is well known or is a new one. Hence, the affordance representation enables the agent to potentially bring a substantial amount of knowledge to new tasks that are to be solved.

One important form of interaction is that of grasping. For a given object, how might an agent come to represent the set of feasible grasps that may be made? Ultimately, one must establish a mapping from perceivable visual and haptic features to a set of parameterized grasping actions (specific positions and orientations for the hand, as well as configurations for the fingers) that are expected to be successful if executed. We would like for these representations to be rooted in an agent's own experiences – either through direct interaction with objects or through observation of other agents' interactions.

In this chapter, we describe two efforts toward addressing this challenge. First, we describe an approach for visually recognizing the 3D orientation of an object. The models of object appearance are based entirely on sequences of image/object pose pairs as the object is being manipulated. The learning algorithm identifies robust descriptions of object appearance from different viewing angles. Second, we introduce a method of identifying descriptions of canonical grasps (that include hand pose and finger configuration) based on observation of a large number of example grasps made by a human teacher. We employ a clustering method in this *hand posture space* that identifies a small number of these canonical grasps. The resulting grasp descriptions can then be used by the agent for planning and execution of grasping actions and for interpreting the grasping actions of other agents.

## 4.2 Learning Models of 3D Object Appearance

One of our ultimate goals is for a robotic agent to learn affordance representations based on experience gathered by looking at an object as the agent manipulates it. In particular, we would like to construct visual models that enable the agent to recognize the object and the angle from which it is being viewed. This interactive approach means that although the agent is able to control many of the conditions in which this experience is gathered, the learning approach must be robust to spurious features in the visual stream, including occlusions by the robot itself and lighting effects such as shadows and specular reflections. The challenge is to discover visual operators that are sensitive to the appearance of the object at some subset of viewing angles (or *aspects*), but that are not "distracted" by these spurious effects. In our approach, individual visual operators recognize the appearance of an object for a subset of viewing aspects. A complete 3D appearance model of an object is captured by identifying a collection of visual operators that cover all possible viewing aspects.

### *4.2.1 Edgel Constellations for Describing 2D Object Appearance*

A visual operator in our case recognizes a specific constellation of oriented *edgels* (129; 45; 191). Edgels are edge image features defined at each pixel, and are described by their orientation in the image and their magnitude. Piater and Grupen (2002) define a *constellation* as a set of edgels that are arranged in some geometric configuration in the 2D image space. This geometric configuration is represented by the relative position and orientation between edgels in a constellation. By construction, a constellation is rotation-invariant in the image plane.

Fig. 4.1 illustrates two constellations that have been identified for two distinct viewing angles of a cup. The constellation that matches the side view (a) captures the top rim of the cup. The constellation that matches the bottom view (b) captures edgels on both the top rim and the bottom of the cup. Within a novel image, a constellation is considered to match if all of the constellation edgels can be found at the correct relative position and orientation. The highest degree of match occurs when the set of edgels in the constellation align perfectly with high-magnitude edges in the query image.

### *4.2.2 Capturing Object Appearance in 3D*

Although the 2D features are invariant under rotation within the image plane, it is clear from Fig. 4.1 that the rotations out of this plane can dramatically alter the appearance of an object. We can represent all possible viewing aspects as the set of points on the unit sphere with the observed object as the center (Fig. 4.2). Imagine a

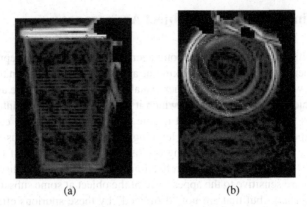

(a)                                    (b)

**Fig. 4.1** Constellations matching a side view (a) and a bottom view (b) of a cup. The constellations have been "painted" on top of the edge magnitude image for each viewing direction. Individual edgels are shown using small circles; the geometric constraints between edgels are shown as dotted lines.

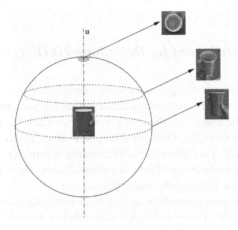

**Fig. 4.2** The aspect sphere of a cup with rotation symmetry about **u**.

camera located at some point on this *aspect sphere*, oriented toward the sphere's origin. This point therefore constrains two orientation DOFs, leaving free the rotation about the camera's axis. The object appearance at a single point can be described by one (or a small number of) edgel constellations. For the case of the cup, one can imagine a unique constellation that only matches a set of views surrounding the top pole of the sphere. As the viewing angle deviates from **u**, the likelihood of observing the constellation can drop quickly. For a constellation that recognizes a "non-polar" aspect, the set of recognized aspects will fall along a circular band on the sphere. This is because the cup's appearance does not change with rotations about the vertical axis As the viewing angle deviates from the center of the band, it becomes less likely that the constellation will be observed.

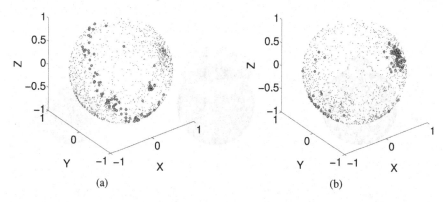

**Fig. 4.3** Constellation matches for (a) the side constellation (of Fig. 1(a)) and (b) the bottom constellation (Fig. 1(b)). Small dots show aspects that correspond to image samples; large dots indicate where the constellation matches. In this case, the major axis of the cup is aligned with the X axis.

Fig. 4.3 illustrates the aspects for which the constellations of Fig. 4.1 are found. The major axis of the cup in these figures falls along the X axis, with the top of the cup at $X = 1$. The constellation that recognizes the side of the cup is found most often along the circle for which $X = 0$ (a). The constellation that recognizes the bottom of the cup is found most often around $X = 1$, but is occasionally found around $X = 1$ (b). This is the case because this particular constellation recognizes pieces of two concentric circles of particular radii, a feature that is sometimes visible from the top of the cup.

How can we compactly represent the set of viewing angles for which a constellation is viewable? Specifically, we would like to capture the likelihood of the aspect given that a particular constellation $C_i$ has been observed: $p(\mathbf{a}\ Obj, C_i)$. (30) proposed the *small circle distribution*, which allows us to describe Gaussian-like distributions on the unit sphere.

$$b(\mathbf{a}\ \tau, \nu, \mu) = \frac{1}{F(\tau, \nu)} e^{\ \tau(\mu^T \mathbf{a}\ \nu)^2}, \tag{4.1}$$

where $\mathbf{a}$, $\mu$ are unit vectors and $\mu$ denotes the mean direction; $\tau$ is a scalar that gives the concentration of the distribution (the higher the $\tau$, the more concentrated the distribution); $\nu$ is a scalar that determines the shape of the distribution; and $F(\tau, \nu)$ is a normalizing term. Note that equation 4.1 obtains a maximum value when $\mu^T \mathbf{a} = \nu$. This set of $\mathbf{a}$'s fall at a fixed angle about $\mu$. By adjusting the parameters of this distribution, we can describe different shapes of clusters on our aspect sphere (Fig. 4.4).

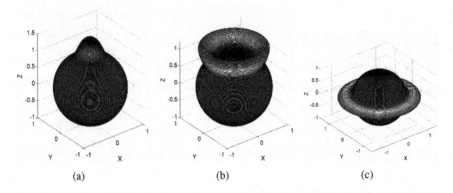

**Fig. 4.4** Gaussian-like distributions on the unit sphere, where $\mu = [0, 0, 1]$. In all cases, the surface radius is $1 + p/(2 \quad \max(p))$, where p is the likelihood at the corresponding aspect. (a) Uniform Gaussian: $\tau = 50$, $\nu = 1.2$; (b) small circle: $\tau = 100$, $\nu = 0.8$; and (c) great circle: $\tau = 100$, $\nu = 0$.

## 4.2.3 Learning Complete 3D Appearance Models

Given a set of image/aspect tuples, the challenge is to discover a set of edgel constellations that cover the entire aspect sphere. Our algorithm is outlined in Fig. 4.5. During the training process, the algorithm samples a single constellation at a time from a specific training image (call this image P). A two-step filtering process is used to determine whether the constellation is *locally robust* and *globally discriminative*. First, the set of images from aspects surrounding P are searched for the constellation. If the constellation describes transient features such as shadows, then it is unlikely to match neighboring images. If this is the case, the constellation is discarded and a new one is sampled. The second filter examines all images in the training set for the constellation. If the degree of match of the constellation distinguishes the neighboring images from most of the remaining training set, then the constellation is considered to be discriminative. Formally, the discriminative power of the constellation is measured using the Kolmogorov-Smirnoff distance (KSD) between the neighboring and complete population of images (130). Should the constellation satisfies both filters, the algorithm then finds the parameters of a probability density function that describes the set of aspects in which the constellation is observed. This training process continues iteratively until the entire set of generated constellations cover most of the training images.

Given a novel image, we would like to accurately estimate the aspect from which it is being viewed. More specifically, assuming that a set of constellations $C_1, C_2, ..., C_N$ are either observed or not in an image, we would like to find the aspect, $a$ that maximizes $p(\mathbf{a} \ Obj, C_1, ..., C_N)$. Making the naïve Bayes assumption, we can estimate this likelihood accordingly:

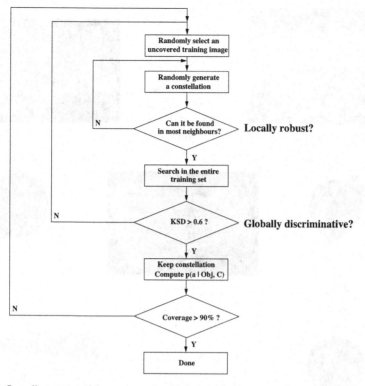

**Fig. 4.5**  Overall structure of the aspect recognition algorithm.

$$p(\mathbf{a} \, Obj, C_1, ..., C_N) = \prod_{i=1}^{N} p(\mathbf{a} \, Obj, C_i). \qquad (4.2)$$

In practice, we make use of a local gradient ascent search with multiple starting locations to identify the maximum likelihood **a**.

## 4.2.4 Data Collection and Preprocessing

In our experiments, each element in the data set is a *tuple* of image and object pose. A Polhemus Patriot (Colchester, VT) is attached to the object so that the 3D position and orientation of the object can be paired with each image. Tuples are gathered continuously as the object is rotated in front of the camera. In all, a data set will contain about 2000 such tuples. We employ an image preprocessing step that identifies a region of interest (ROI) that ideally contains only the object. The stationary background and skin-colored objects are first subtracted from the image. The ROI is then selected to encompass a large, contiguous set of the remaining

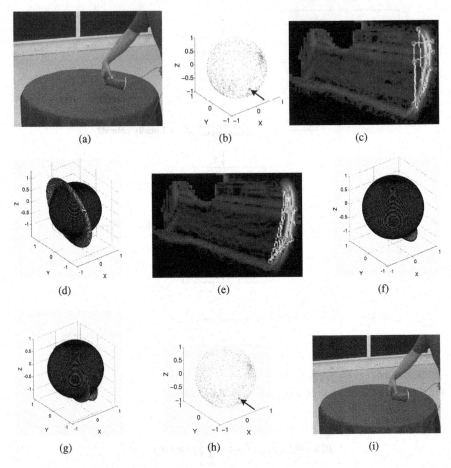

**Fig. 4.6** Example aspect recognition. (a) a testing image of the cup, (b) the true aspect from which the object is observed, (c) match of constellation 1 to the test image, and (d) $p(\mathbf{a}\ cup, C_1)$. (e) match of constellation 2 to the test image, (f) $p(\mathbf{a}\ cup, C_2)$, (g) $p(\mathbf{a}\ cup, C_1, C_2)$, (h) the maximum likelihood aspect, and (i) the nearest training image that corresponds to this aspect.

pixels within the middle of the image frame. In practice, the ROI contains the object in excess of 99% of the images.

Fig. 4.6 illustrates the recognition process. An independent testing image of the cup is shown in Fig. 4.6(a) and the corresponding (true) aspect is shown on the aspect sphere in Fig. 4.6(b). For this particular image, two constellations are observed (Fig. 4.6(c) and Fig. 4.6(e)). The density functions corresponding to these two constellations are shown in Fig. 4.6(d) and Fig. 4.6(f). The combined density function (Eq. 4.2) is shown in Fig. 4.6(g). The maximum likelihood aspect is 0.31 from the true aspect (Fig. 4.6h).

## *4.2.5 Experimental Results*

Both symmetric (a cup and a block) and asymmetric (a mug and a spray bottle) objects are used in the experiment. For each object, about 2000 sample image/aspect tuples are taken uniformly in order to cover the aspect sphere as well as possible. For each object, we performed 10 independent experiments. For each experiment, a different set of 100 samples are randomly selected and reserved as the test data set; the remaining samples are used as training data. Error is measured for each test image as the angle between the estimated and true aspects, down to the symmetry of the object. When there are multiple estimated aspects, the mean of error is calculated for a single test image. We report the mean error over 100 images and 10 experiments. We compare the proposed approach with one in which no filtering is performed (the "unfiltered method") and with a method that guesses aspects randomly.

Both the filtered and unfiltered methods cover 3556 out of 4000 testing images (of 4 objects, 100 test images and 10 experiments). No constellations are found in the remaining test images. The aspect estimate error histogram for the three methods is shown in Figure 4.7a. These histograms include errors from all ten experiments and four objects. The filtered and unfiltered methods decrease exponentially with increasing error. However, the filtered method is biased more toward lower errors. The mean error for the random method is substantially higher than either of the other two methods.

The mean errors and standard deviations for each object are shown in Fig. 4.7b. For both methods, we can see that the errors for the spray bottle are relatively large compared to that for the other objects. The reason is that the shape and texture of the spray bottle are more complex than the other objects. As a result, many constellations often match to a high degree with the texture of the labels, even though they are not originally generated from those regions.

We can also see that filtering is a benefit, especially for the more complicated objects. For the simple objects, sampled constellations for a particular aspect are often very similar to each other. Hence, the filtering step does not make any practical distinctions between different constellations. As the objects become more complicated, such as with the spray bottle, a particular aspect will give rise to a set of rather different constellations. Hence, the filtering step is able to make meaningful distinctions between these constellations.

The performance difference between the filtered and unfiltered methods is significant for all four objects by a two-tail, paired t-test (block: $p < 0.05$; cup: $p < 10^{3}$; mug: $p < 10^{4}$; spray bottle: $p < 10^{4}$). We should also note that the random guess method does not perform as poorly as one might expect. This is because these errors have also been adjusted according to the symmetric properties of the objects.

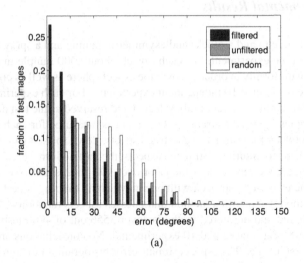

(a)

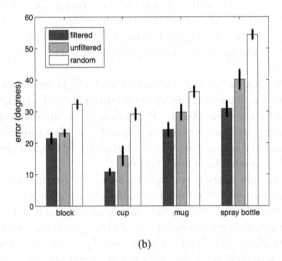

(b)

**Fig. 4.7** (a) Histogram of aspect estimate errors made by the three methods (all objects), and (b) aspect estimation errors for each method and object.

## 4.3 Learning Canonical Grasps for Objects

Once an object has been visually identified and localized in space, how can an agent describe the set of possible grasping actions that may be taken? Because the set of possible actions will ultimately be used for planning and for exploratory learning, we are motivated to make this set as small as possible so as to reduce the complexity of either search process. One approach to constructing this representation is to begin with a large set of successful example grasps and then to cluster them into a small set of canonical grasps. This set of examples could be derived from manipulation sequences produced by the agent itself, or by a human acting directly on the manipulated object or acting through the agent via teleoperation. Our focus to date has been on these two human-driven methods.

We describe each example grasp with the following: 1) position of the hand in an object-centered coordinate frame, 2) orientation of the hand, and (in some cases) 3) the joint angles of the fingers of the hand. Clustering is performed using a *mixture of probability density functions* approach, in which each cluster corresponds to a canonical hand configuration that describes all three of these components (de Granville et al., 2006, 2009, and submitted; de Granville, 2008). Below, we detail each of these steps and then show that this method can successfully identify meaningful clusters from teleoperation experiments performed using NASA's humanoid robot *Robonaut*.

### 4.3.1 Modeling Hand Orientation

Unit quaternions are a natural representation of 3D orientation because they comprise a proper metric space, a property that allows us to compute measures of similarity between pairs of orientations. Here, an orientation is represented as a point on the surface of a 4D unit hypersphere. This representation is also antipodally symmetric: pairs of points that fall on opposite poles represent the same 3D orientation. The Dimroth-Watson distribution captures a Gaussian-like shape on the unit hypersphere, while explicitly acknowledging this symmetry (110; 138). The probability density function for this distribution is as follows:

$$f(\mathbf{q} \mid \mathbf{u}, k) = F(k) e^{k(\mathbf{q}^T \mathbf{u})^2}, \tag{4.3}$$

where $\mathbf{q} \in \mathbb{R}^4$ represents a unit quaternion, $\mathbf{u} \in \mathbb{R}^4$ is a unit vector that represents the "mean" rotation, $k \geq 0$ is a concentration parameter, and $F(k)$ is a normalization term. Note that $\mathbf{q}^T \mathbf{u} = \cos \theta$, where $\theta$ is the angle between $\mathbf{q}$ and $\mathbf{u}$. Hence, density is maximal when $\mathbf{q}$ and $\mathbf{u}$ are aligned, and decreases exponentially as $\cos \theta$ decreases. When $k = 0$, the distribution is uniform across all rotations; as $k$ increases, the distribution concentrates about $\mathbf{u}$. Fig. 4.8(a) shows a 3D visualization of the Dimroth-Watson distribution, and highlights its Gaussian-like characteristics. The high density peaks correspond to $\mathbf{u}$ and $-\mathbf{u}$.

A second cluster type of interest corresponds to the case in which an object exhibits a rotational symmetry. For example, an object such as a cylinder can be approached from any orientation in which the palm of the hand is parallel to the planar face of the cylinder. In this case, hand orientation is constrained in two dimensions, but the third is unconstrained. This set of hand orientations corresponds to an arbitrary rotation about a fixed axis, and is described by a great circle (or girdle) on the 4D hypersphere. We model this set using a generalization of the Dimroth-Watson distribution that was suggested by (144). The probability density function is as

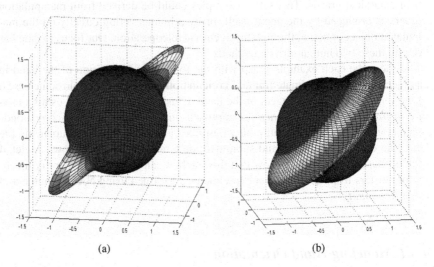

**Fig. 4.8** Three dimensional representations of the Dimroth-Watson (a) and girdle (b) distributions on S2. In both cases, the surface radius is $1 + p$, where $p$ is the probability density at the corresponding orientation

follows:

$$\bar{f}(\mathbf{q}\ \mathbf{u}_1, \mathbf{u}_2, k) = \bar{F}(k) e^{k\left[(\mathbf{q}^T\mathbf{u}_1)^2 + (\mathbf{q}^T\mathbf{u}_2)^2\right]}, \tag{4.4}$$

where $\mathbf{u}_1 \quad \mathbb{R}^4$ and $\mathbf{u}_2 \quad \mathbb{R}^4$ are orthogonal unit vectors that determine the great circle, and $\bar{F}(k)$ is the corresponding normalization term. Fig. 4.8(b) illustrates the girdle distribution on the 3D unit sphere. First, note that all points on the great circle are assigned maximal density. This corresponds to the set of points for which $\left(\mathbf{q}^T\mathbf{u}_1\right)^2 + \left(\mathbf{q}^T\mathbf{u}_2\right)^2 = 1$. However, as the angle between $\mathbf{q}$ and the closest point on the circle increases, the density decreases exponentially.

For a given set of observations, the parameters of the Dimroth-Watson and girdle distributions are estimated using maximum likelihood estimation (MLE). The axes of the distribution are derived from the sample covariance matrix, $\Lambda \quad \mathbb{R}^4 \ ^4$:

$$\Lambda = \frac{\sum_{i=1}^{N} \mathbf{q}_i \mathbf{q}_i^T}{N}, \tag{4.5}$$

where $\mathbf{q}_i$ is the orientation of the $i$th sample, and $N$ is the total number of samples. The MLE of $\mathbf{u}$ is parallel to the first eigenvector of $\Lambda$ (110; 138). The orthogonal vectors $\mathbf{u}_1$ and $\mathbf{u}_2$ span the same space as the first and second eigenvectors of $\Lambda$ (144).

For the Dimroth-Watson distribution, the MLE of the concentration parameter, $k$, uniquely satisfies the following (see (78) for the derivation):

$$G(k) = \frac{F'(k)}{F(k)} = \frac{\sum_{i=1}^{N} \left(\mathbf{q}_i^T \mathbf{u}\right)^2}{N}. \qquad (4.6)$$

In the case of the girdle distribution, the MLE of $k$ uniquely satisfies:

$$\bar{G}(k) = \frac{\bar{F}'(k)}{\bar{F}(k)} = \frac{\sum_{i=1}^{N} \left[ \left(\mathbf{q}_i^T \mathbf{u}_1\right)^2 + \left(\mathbf{q}_i^T \mathbf{u}_2\right)^2 \right]}{N}. \qquad (4.7)$$

For computational efficiency, we approximate $G^{-1}()$ and $\bar{G}^{-1}()$ when solving for $k$. This approximation is discussed in detail by (78).

### 4.3.2 Modeling Hand Position

The position of the hand is represented as a 3D vector in Cartesian space. We choose to model position using a Gaussian distribution:

$$p(\mathbf{x}|\mu, \Sigma) = \frac{1}{(2\pi)^{\frac{d}{2}} |\Sigma|^{\frac{1}{2}}} e^{-\frac{1}{2}(\mathbf{x}-\mu)^T \Sigma^{-1}(\mathbf{x}-\mu)}. \qquad (4.8)$$

Here, $\mathbf{x} \in \mathbb{R}^d$ denotes a point in a $d$ dimensional Cartesian space, while $\mu \in \mathbb{R}^d$ and $\Sigma \in \mathbb{R}^{d \times d}$ correspond to the mean vector and covariance matrix of the Gaussian distribution. For our purposes, $d = 3$, $\mu$ describes the mean position of the hand, and $\Sigma$ captures covariance in hand position.

### 4.3.3 Modeling Finger Posture

Humanoid robots such as Robonaut typically have many degrees of freedom (DOF) available to perform manipulation tasks. For example, each of Robonaut's hands has 12 DOF: three for the thumb, index, and middle fingers; one for the ring and pinkie fingers; and one for the palm (13). Incorporating finger configurations into our clustering algorithm is a key step to constructing more complete grasp affordance representations. One possible approach to this problem is to learn clusters using the full dimensionality of the robot's end-effector. However, hands with a large number of joints can be difficult to model because an increasingly large number of training examples is needed to adequately sample a space as more and more dimensions are added to it.

One question is whether or not all of the available DOFs of the hand are even necessary to accurately model the finger configurations used for grasping. For example, when executing a power grasp, the fingers tend to flex in unison. This means that there is a strong correlation between the distal and proximal joints of each finger, as well as a correlation across fingers. (148) and (43) present an approach that takes advantage of such correlations through the notion of an *eigengrasp*. The eigengrasps of a hand comprise a set of basis vectors in the joint space of the hand. Linear combinations of a small number these eigengrasps can be used to approximate the finger configurations used when grasping.

More formally, let $\mathbf{p}$ $\mathbb{R}^d$ be a column vector of joint angles describing the finger configuration of a robot's end-effector, and $\mathbf{V}$ $\mathbb{R}^{d\ d}$ constitute a basis for the vector space of which $\mathbf{p}$ is a member. The columns of $\mathbf{V}$ represent directions in the joint space of the hand (the eigengrasps), and are ordered from those that capture the most variance in finger configuration to the smallest (i.e., from the largest corresponding eigenvalue to the smallest). Linear combinations of the columns of $\mathbf{V}$ can be used to represent any possible pose for the fingers of the robot's hand:

$$\mathbf{p} = \sum_{i=1}^{d} a_i \mathbf{v}_i = \mathbf{Va}. \tag{4.9}$$

Here, $\mathbf{v}_i$ $\mathbb{R}^d$ is the $i$'th column of $\mathbf{V}$, $\mathbf{a}$ $\mathbb{R}^d$ is a column vector of coefficients, and $a_i$ $\mathbb{R}$ is element $i$ of the vector $\mathbf{a}$.

Because there may be a large number of joints in the robot's hand, the configuration of the fingers may be approximated by using a small number ($K$) of eigengrasps:

$$\mathbf{p} = \sum_{i=1}^{K} a_i \mathbf{v}_i = \hat{\mathbf{V}}\hat{\mathbf{a}}, \tag{4.10}$$

where $\hat{\mathbf{V}} = \begin{bmatrix} \mathbf{v}_1 & \mathbf{v}_2 & ... & \mathbf{v}_K \end{bmatrix}$, and $\hat{\mathbf{a}} = \begin{bmatrix} a_1 & a_2 & ... & a_K \end{bmatrix}^T$. Given a finger configuration $\mathbf{p}$ and a subset of the eigengrasps $\hat{\mathbf{V}}$, a low dimensional representation of $\mathbf{p}$ is obtained by solving the system of linear equations in 4.10 for $\hat{\mathbf{a}}$.

We compute the set of eigengrasps using samples of the joint angle vector as a teleoperator grasps a set of objects. Let $\mathbf{P}$ $\mathbb{R}^{d\ N}$ be the set of finger configurations resulting from the human demonstration, where $N$ denotes the number of examples. The eigengrasps are determined by computing the eigenvectors of $\mathbf{P}$'s covariance matrix (80).

### 4.3.4 Modeling Mixtures of Hand Postures

We model a grasp using a joint distribution defined over hand pose and the finger posture. Specifically:

$$g(\mathbf{x}, \mathbf{q}, \hat{\mathbf{a}}\ \theta) = p(\mathbf{x}\ \theta_p) f(\mathbf{q}\ \theta_f) p(\hat{\mathbf{a}}\ \theta_h), \tag{4.11}$$

and

$$\bar{g}(\mathbf{x}, \mathbf{q}, \hat{\mathbf{a}}\ \bar{\theta}) = p(\mathbf{x}\ \theta_p) \bar{f}(\mathbf{q}\ \theta_{\bar{f}}) p(\hat{\mathbf{a}}\ \theta_h). \tag{4.12}$$

Here, $p(\hat{\mathbf{a}}\ \theta_h)$ is a multivariate Gaussian distribution over $K$ dimensions. We assume that hand position, hand orientation, and finger configuration are conditionally independent given a cluster.

An individual hand posture distribution can capture a single cluster of points, but a set of grasps is typically fit best by multiple clusters. Furthermore, the use of multiple clusters captures any covariance that may exist between the position and orientation of the hand when grasping a particular object. We therefore employ a mixture model-based approach. Here, the density function of the mixture, $h()$, is defined as:

$$h(\mathbf{x}, \mathbf{q}\ \Psi) = \sum_{j=1}^{M} w_j c_j(\mathbf{x}, \mathbf{q}\ \theta_j), \qquad (4.13)$$

$$\Psi = (w_1, ..., w_M, \theta_1, ..., \theta_M), \qquad (4.14)$$

and

$$\sum_{j=1}^{M} w_j = 1, \qquad (4.15)$$

where $M$ denotes the number of component densities, and $c_j$ is one of the two density functions describing hand pose ($g()$ or $\bar{g}()$). Each element of the mixture represents a single cluster of points, and is weighted by $w_j$. Estimation of the parameters of the individual clusters and the cluster weight variables is accomplished using the Expectation Maximization (EM) algorithm (57).

For a given set of observations, it is unclear *a priori* how many or of what type of cluster is appropriate. Our approach is to construct all possible mixtures that have a maximum of $M$ clusters (we choose $M = 10$) and to choose the mixture that best matches the observations. For this purpose, we make use of the Integrated Completed Likelihood (ICL) criterion (29) to evaluate and order the different mixture models. Like the Bayesian Information Criterion, ICL prefers models that explain the training data, but punishes more complex models. In addition, ICL punishes models in which clusters overlap one-another. These features help to select models that describe a large number of grasps with a small number of clusters.

Because the EM algorithm is a gradient ascent method in a likelihood space containing many local maxima, each candidate mixture model was fit a total of $\Omega$ different times using the available training data (for our purposes, $\Omega = 80$). For a given mixture, this ensures that a variety of different initializations for the EM algorithm are explored. The model that performs best on the first validation set according to ICL is subsequently evaluated and compared with other mixtures using the second validation set (again using ICL).

Due to our data collection procedure, some samples do not correspond to quality grasps, and instead correspond to transitions between grasps. It is desirable that our clustering algorithm be robust to this form of noise. However, when a large enough number of mixture components is allowed, the EM algorithm tends to allocate one or more clusters to this small number of "outlier" samples. We explicitly discard these mixture models when an individual cluster covers a very small percentage of the samples (indicated by a small magnitude cluster weight parameter, $w_j$). In particular, a model is discarded when:

$$\frac{\max_j(w_j)}{\min_j(w_j)} \quad \lambda, \tag{4.16}$$

where $\lambda$ is a threshold. For our experiments, we choose $\lambda = 5$ because it tends to result in the selection of high quality, compact models. Of the models that have not been removed by this filter step, the one with the best ICL measure on the second validation set is considered to be the best explanation of the observed data set.

### 4.3.5 Data Collection

The human teleoperator is able to control Robonaut's many degrees of freedom with a virtual reality-like helmet and a data glove equipped with a Polhemus sensor (13). In addition to articulating Robonaut's neck, the helmet provides visual feedback from the environment to the teleoperator. The arms and hands of the robot are commanded by tracking the movements of the human's wrists and fingers, and performing a mapping from human motion to robot motion.

Each trial consists of the human teacher haptically exploring an object for approximately 15 minutes. The object is located in a fixed pose relative to the robot. To maximize the number of quality samples collected, different grasping strategies may be employed by the teleoperator based on the local geometry of the object. For example, when grasping larger surfaces, a sliding motion in conjunction with a fixed finger configuration is used. This ensures that the feasible positions and orientations of the hand are collected in a timely manner. In contrast, the teleoperator repeatedly opens and closes the robot's hand when grasping small surfaces. This strategy forces hand pose to vary even though the hand may not be able to slide along the local surface.

When compared with the data collected during direct observation of a human performing grasping actions (78; 54), the robot teleoperation experience tends to contain larger amounts of noise. Robonaut's arm motions are slower and less fluid under human control. Hence, the hand posture samples contain a large number of cases in which the hand is not in contact with the object. To alleviate this problem the transitions are removed manually by identifying the time intervals in which they occur.

### 4.3.6 Experimental Results

To demonstrate the effects of incorporating finger configuration into the grasp learning process, a number of experiments are performed. First, the eigengrasps are learned based on experience that is generated by the human teleoperator. A num-

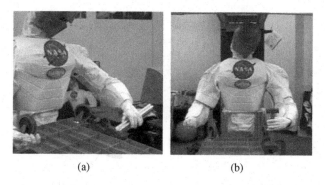

**Fig. 4.9**  The set of objects used in the Robonaut clustering experiments. (a) Handrail; (b) Hammer.

ber of different objects are used to ensure a reasonable sampling of the finger configurations. Due to invalid sensor data, seven of the finger joints are ignored. This means that the number of effective degrees of freedom in Robonaut's hand has been reduced from twelve to five. Of these remaining five degrees of freedom, approximately 98% of the variance can be explained by the first three principal components. This, in conjunction with the ability to visualize the resulting low dimensional representation of finger configuration, led to the use of only the first three eigengrasps (i.e., $K = 3$).

### 4.3.6.1 Handrail

Figs. 4.10(a,c,e) show the training examples for the handrail object. Panel (a) shows the 3D position of the hand throughout the course of the experiment, while panel (b) provides a visualization of the corresponding hand orientations. Orientation of the hand is represented as a single point on the surface of the unit sphere: imagine that the object is located at the origin of the sphere; the point on the surface of the sphere corresponds to the intersection of the palm with the sphere. Note that this visualization technique aliases the set of rotations about the line perpendicular to the palm. In both panels (a) and (c), the major axis of the handrail is located along the $X$ axis, with the grasped end at $X = \quad 60$ in the position space and at $X = \quad 1$ in the orientation space. Finally, panel (c) shows the finger configurations projected into the eigengrasp space.

A total of five clusters were learned for the handrail object: two that correspond to an overhand reach in which the handrail is approached from the top, two for the underhand configuration, and one for the side approach. The learned position clusters are shown in Fig. 4.10(b) as first standard deviation ellipsoids of the Gaussian distribution. The orientation component of these clusters is represented using a Dimroth-Watson distribution and is show in panel (d). The mean orientation is indicated using the line segment emanating from the center of the sphere. Clusters 1 and

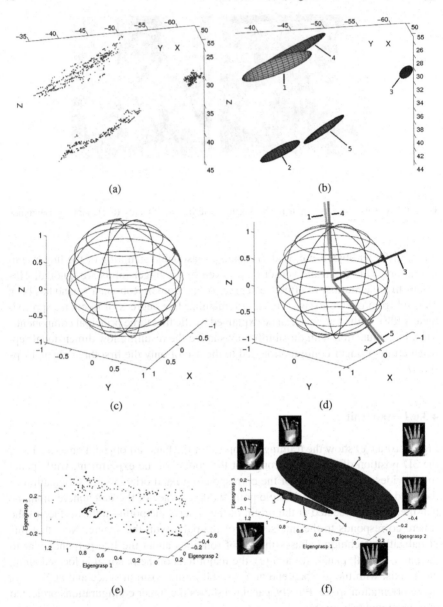

**Fig. 4.10** The training examples and learned affordance model for the handrail. (a) The position of the hand; (b) The position component of the learned affordance model; (c) The orientation of the hand; (d) The orientation component of the learned affordance model; (e) The finger configuration of the hand; (f) The finger configuration component of the learned affordance model.

4 correspond to the top approach and show an elongation in the position component along the $X$ axis. This elongation encodes the fact that the top approach results in grasps at many points along the length of the handrail. Likewise, clusters 2 and 5 correspond to the underhand approach (with the palm up) and are also elongated along the handrail. Cluster 3 corresponds to the side approach to the handrail. The demonstrated variation in hand position was very small, and is reflected in the size of the ellipsoid.

Panel (f) shows the learned eigengrasp clusters. Each corner of the bounding box provides a visualization of the mapping that occurs between the low dimensional representation of finger configuration and each joint of the robot's hand. Notice that variation along the first eigengrasp corresponds to flexion of the index and middle fingers, while variation along the second eigengrasp causes adduction and abduction of the thumb. However, variation along the third eigengrasp does not affect the configuration of the fingers significantly, only affecting the flexion of the most distal joints of the index and middle fingers. Also, note that the ring and pinkie fingers remain in their extended configurations. (these are among the degrees of freedom for which no data were recorded).

Turning to the learned eigengrasp clusters, notice that the ellipsoids 1 and 2 are in the same region of the finger configuration space even though they correspond to grasp approaches from above and below the handrail. Because the same sliding technique was employed by the teleoperator when demonstrating these grasps, the hand had a similar shape for each approach. However, for cluster 2 there is more variation in finger configuration, which is indicated by the elongation of ellipsoid 2. In contrast, the hand was continually opened and closed when the side approach was used to grasp the handrail. This is evident by comparing the hand shapes that correspond to points on opposite ends of ellipsoid 3's major axis. On the right end of the figure, the hand is in an open configuration, but on the left end the middle and index fingers are flexed considerably. Also, notice that ellipsoid 3 is separated from the other eigengrasp clusters, which highlights the different hand shapes used when grasping the handrail from above and below versus from the side.

#### 4.3.6.2 Hammer

The example grasps demonstrated by the human teleoperator and the learned grasp affordance model for the hammer are shown in Fig. 4.11. In this case five clusters were learned: cluster 1 represents grasps when approaching from above the hammer's head. The orientation of this cluster is represented using a girdle distribution, as indicated by the circle on the surface of the sphere in panel (d). In our visualization, the points along the circle correspond to the orientations of maximum density. For the case of cluster 1, we would have expected the use of a Dimroth-Watson distribution because there is little variation about the points corresponding to the top approach. However, what variation there is falls along a narrow arc that is best captured by the girdle.

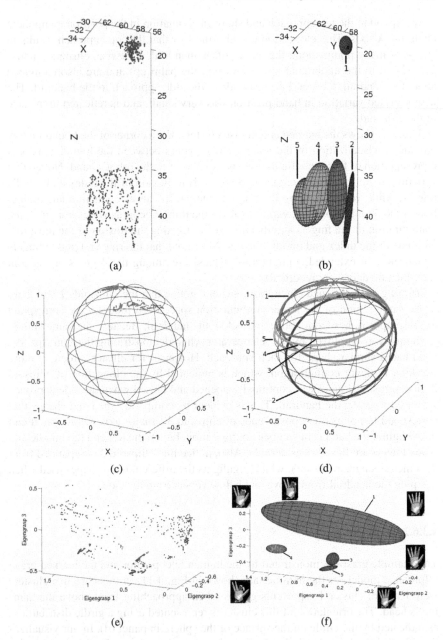

**Fig. 4.11** The training examples and learned affordance model for the hammer. (a) The position of the hand; (b) The position component of the learned affordance model; (c) The orientation of the hand; (d) The orientation component of the learned affordance model; (e) The finger configuration of the hand; (f) The finger configuration component of the learned affordance model.

The remaining clusters (2–5) capture grasps along the handle of the hammer when approaching from the side. Girdle distributions were selected to model the orientation of the hand for the side approach. While this is encouraging, the algorithm learned four clusters instead of one. This is most likely due to the spatially distinct hand positions used to grasp the hammer's handle.

The finger configurations used by the teleoperator to grasp the hammer's handle were very different than those used to grasp the hammer's head. When approaching the object from the side, power grasps that maximized the contact surface area between the hand and the handle were more likely to be used. Conversely, precision grasps that mainly used the finger tips were employed when grasping the head of the hammer. These differences in hand shape can be seen in Fig. 4.11(f). Ellipsoid 1 represents the finger configurations used to grasp the hammer from above. The large volume of the ellipsoid is due to the exploration strategy employed by the teleoperator: the hand was continually opened and closed on this portion of the object. Hence, there was a large variance in finger configuration. Also, notice that ellipsoids 2–5 are spatially distinct from eigengrasp cluster 1.

## 4.4 Discussion

In this chapter, we presented several steps toward robot learning of affordance representations in support of grasping activities. Affordances provide a means of mapping sensory information, including vision, into a small set of grasping actions that are possible with the object being viewed. Key to this representation is the fact that it captures the specific interaction between the object and the agent. The ability to learn these representations automatically will be important as we begin to field robots in unstructured environments and expect them to perform a wide range of manipulation tasks.

Given a sequence of tuples consisting of an image and an object pose, our algorithm learns 3D appearance models for objects. In particular, the algorithm identifies visual operators that are robust to spurious image features such as object occlusions and shadows. Visual operators are implemented as edgel constellations that describe a specific geometrical relationship between a set of small edges. The 3D appearance of an object is captured by compactly describing the set of viewing angles from which each image feature is viewable. When a novel image is presented, the set of observed features can then be used to estimate the most likely viewing angle of the object. Ultimately, we will estimate the complete pose of the object, which, in turn, can be used for planning and executing grasping actions.

In more recent work, we have begun to make use of scale-invariant feature transform (SIFT) image features in place of edgels (109). This method is showing promise in addressing image scale issues, improving the computational efficiency of identifying features, and increasing the accuracy of the viewing angle estimates. One of the challenges in using such an approach is that of pruning the set of primitive features that arise from such a large database of images that exhibit very similar appearance. In addition, we are now making use of a particle-based approach for describing the density functions. This approach is helping to address the overfitting issues that can arise with mixture-type models and allows us to capture irregular shapes in the density functions.

The second component of our algorithm uses tuples of object pose and hand posture to construct a small menu of grasps that are possible with the object. These compact representations are constructed from many example grasps made by clustering the hand posture examples. This property enables the use of the affordance as a way to access "primitives" in higher-level activities, including planning, learning, and the recognition of motor actions by other agents (64; 34). In particular, the hand posture clusters that have been learned map directly onto resolved-rate controllers that can bring a robot hand to a specific position and orientation relative to the object. Note that this control step assumes that haptic exploration methods are available to refine the grasps once the hand has approached the object (Coelho and Grupen 1997; Platt et al., 2002, 2003; Platt 2006; Wang et al., 2007).

Our approach to date has assumed an intermediate representation between vision and grasp that is rooted in the individual objects. However, we would ultimately like for this representation to be able to generalize across objects. This step will be important as the robotic agent is faced with objects with which it has little to no prior experience. Our approach is to identify canonical grasps that routinely co-occur with particular visual features. When two or more objects have components that share a common shape, and hence common visual features, it is likely that similar hand postures will be used to grasp these components.

The affordance representation captures the syntax of grasping (i.e., what grasps are possible for a given object), and does not take into account the semantics of grasping (how an object is to be used in the larger context of a task). This distinction, which is drawn by Gibson, is a critical one for a learning agent. When a new task is presented, the syntax of interacting with a specific object can be readily accessed and used. The learning agent is then left with the problem of selecting from a small menu of possible grasping actions to solve the new task. This abstraction can have important implications for the agent quickly learning to perform in these novel situations.

**Acknowledgements** This research was supported by NSF/CISE/REU award #0453545 and by NASA EPSCoR grant #7800. We would also like to thank Mike Goza, Myron A. Diftler, and William Bluethmann at the Dexterous Robotics Laboratory at NASA's Johnson Space Center for their support in providing access to Robonaut. We also thank Brian Watson for his contributions to this work.

# Chapter 5
# Intelligent Robotics for Assistive Healthcare and Therapy

Ayanna M. Howard, Sekou Remy, Chung Hyuk Park, Hae Won Park, and
Douglas Brooks

**Abstract** In this chapter, we discuss research methodologies focused on developing intelligent robotics for healthcare applications. In the first section, we present methods in which robot assistants learn to perform daily tasks by generalizing instructions provided through human teleoperation. In the second section, we discuss research that integrates robots and toys in order to institute interactive play behavior for child therapy and education. The last section presents methods for using robots to assist in physical therapy scenarios based on therapist-patient observation. We finally conclude with a discussion on future work in the area of robotics for assistive healthcare and therapy.

Ayanna M. Howard
Human-Automation Systems (HumAnS) Lab, Center for Healthcare Robotics@Health Systems Institute, Georgia Institute of Technology, Atlanta, GA 30332, e-mail: ayanna@ece.gatech.edu

Sekou Remy
Human-Automation Systems (HumAnS) Lab, Center for Healthcare Robotics@Health Systems Institute, Georgia Institute of Technology, Atlanta, GA 30332, e-mail: sekou@ece.gatech.edu

Chung Hyuk Park
Human-Automation Systems (HumAnS) Lab, Center for Healthcare Robotics@Health Systems Institute, Georgia Institute of Technology, Atlanta, GA 30332, e-mail: chunghyuk@ece.gatech.edu

Hae Won Park
Human-Automation Systems (HumAnS) Lab, Center for Healthcare Robotics@Health Systems Institute, Georgia Institute of Technology, Atlanta, GA 30332, e-mail: hindol21@ece.gatech.edu

Douglas Brooks
Human-Automation Systems (HumAnS) Lab, Center for Healthcare Robotics@Health Systems Institute, Georgia Institute of Technology, Atlanta, GA 30332, e-mail: bdouglas8@gatech.edu@ece.gatech.edu

## 5.1 Introduction

Ten years ago, research in robotic grasping of deformable objects resulted in the development of a generalized approach for handling of 3-D deformable objects in which prior knowledge of object attributes was not required (86). The motivation of the method was to enable robot grasping of a large class of object types (e.g. asymmetric, nonhomogeneous, nonlinear object types) that are typically found in a hospital environment. The methodology developed was composed of two main threads – determining deformation characteristics for a non-rigid object represented by a physically-based model and utilizing the characteristics to learn the minimum grasping forces required for manipulating such objects.

For determining deformation characteristics, a model was derived by discretizing the deformable object into a network of interconnected particles, springs, and damping elements. By assuming quasi-static system characteristics, nonlinear partial differential equations were used to model the particle motion of the deformable object and calculate the deformation characteristics. For calculating minimum lifting force, an iterative lifting method was developed in which the robotic manipulation system interactively learned by lifting the object with iterative measurements of force. Once the deformation characteristics and the associated lifting force term were determined, they were used as training input into a neural network for generalizing future manipulation operations.

The major conclusion arising from this prior research was the realization that manipulation of *everyday objects* is difficult. Real-world manipulation in the healthcare domain must occur in dynamic and complex environments based on a wide range of tasks that evolve over time. Humans have learned to deal with these difficulties by improving their abilities through the accrual of experiences. It makes logical sense then to develop methods to transfer the experiences of humans to robots in order to endow them with the ability to operate effectively in these environments. Although our fundamental goal is to develop robots that can safely and effectively provide manipulative assistance in the rehabilitation, care, and education of others, we believe this process starts by addressing fundamental problems in knowledge transfer between everyday human users (whether the patient or caregiver) and robots. As such, in this chapter we discuss research methodologies focused on the first steps in developing intelligent robotics for healthcare applications through knowledge transfer. In the first section, we present methods in robot learning in which robot assistants learn to perform new tasks from human direction provided via a haptic interface. In the second section, we discuss research that integrates robots and toys in order to institute interactive play behavior for child therapy and education. The last section presents methods for using robots to assist in physical therapy scenarios based on therapist-patient observation. We finally conclude with a discussion on future work in the area of robotics for assistive healthcare and therapy.

## 5.2 Activities of Daily Living: Robot Learning from Human Teleoperation

Based on the Americans with Disabilities report (2002), 10.7 million Americans (over age 6) need assistance with one or more activities of daily living. Through teleoperation, the process that permits a human operator to directly guide a robotic device, robots can provide remote caregivers with the ability to help individuals with certain activities of daily living while in the home. By coupling teleoperation with robots endowed with manipulation capability, robots can aid caregivers and provide assistance in the home, even without physically being collocated with the healthcare provider. The success of teleoperated assistive robots though depends on how robustly a caregiver can provide instruction and how easily key operations can be automated by the robot in order to increase the reliability of task performance.

In these teleoperation scenarios, the human ultimately controls the operation of the robot by using the robots sensory information for task achievement. It is almost inevitable that the trainers of these assistive robots will not have equivalent technical expertise as the designers of these robots. The human operators will have to be provided with interfaces to control and modify the behaviors of the robots such that they can be optimized for the home environment in which they are deployed. The challenge that must therefore be addressed is to devise a scheme that permits transfer of task knowledge and control of a robot while compensating for errors derived from inexperienced users, i.e. human operators that will most likely not have the time, expertise or exposure to the robot platform to which they are providing instruction.

The main theme of our method for robot learning via human teleoperation is to put the human in the learning loop (Figure 5.1). The advantages in this type of teleoperative control cycle is that a human operator, whether a technical expert or non-expert, can provide much more sophisticated decision making and proper reactions to the situations than a densely programmed robot, upon encountering unexpected conditions in the home environment.

### 5.2.1 Divided Force Guidance for Haptic Feedback

A classical method found in path-planning applications is the potential field method in which calculated forces are used to repel a robot away from obstacles and attract a robot toward a designated goal location. Once these potential fields are calculated, the robot can theoretically navigate to a final goal position within an obstacle-strewn environment by transitioning along the force vectors. For robot learning via human teleoperation, the goal of the haptic feedback system is to assist the human teacher in implementing a new task through teleoperation, such that the control signals provided by the user are not strongly dependant on operator expertise. Based on the concept of the potential field method, we have developed the divided force guidance approach (124) to provide haptic force feedback to the user. In this method,

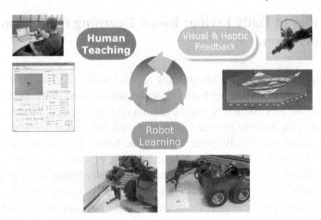

**Fig. 5.1** Description of putting the 'human-in-the-loop' for robot learning. Human teacher controls the robot via teleoperation, while being aided by visual and haptic feedback. The robot student then learns a task from the human operator's guidance.

attractive forces are generated as the user moves toward a given goal and repelling forces are generated when the user moves away from the goal. Diverging from the potential field method, the divided force guidance approach adjusts for both differences in object distances as well as object size. Of most importance, is to account for the subtle differences that are typically found in the potential field method (such as small forces that might be generated when a robot is close to the goal). These forces must be emphasized since too small of a force generated in the haptic device will go unnoticed by the human user. This is accomplished by using vision-based methods to calculate these forces (123). By combining haptics with vision and tele-operation, we provide a more effective means of tele-operative control, such that the user does not require extensive training before interacting with the robot. With this guided process of teleoperation, robot task learning can then be performed.

### 5.2.2 Learning through Haptically Guided Manipulation

Learning of a new task via teleoperation involves capturing all relevant control signals necessary to execute the task once interaction with the human is complete. For task performance, sensing of the target object and understanding the robots action relative to the object are required. Since these reactive cognitions take place over time, we focus on the temporal sequences of sensing and actuation data. Regarding the learning algorithm, we use recurrent artificial neural networks (ANN) and the backpropagation with momentum factor training method (127). ANN is a method of supervised learning, having layers of nodes (neurons) and interconnections with weights forming a network, which can be trained statistically to minimize the error using a gradient descent algorithm. The momentum term is used to solve the local minima problem. Our method for learning execution of a new task focuses on building the neural network by considering actions of the end-effector in the Cartesian space (Figure 5.2). Inputs into the ANN include object position and size (extracted from a visual image stream), end-effector position values, one hidden layer, and one output associated with the differential value for the end-effector. With this neural network structure, we aim to decouple dependencies on the kinematic chain of the manipulator from the learning loop.

**Fig. 5.2** Artificial neural network (ANN) structure of learning for Cartesian space. The ANN inputs are derived from the camera sensor and current position of the end-effector. Outputs are incremental values for x-,y-,z-coordinates for the next movement.

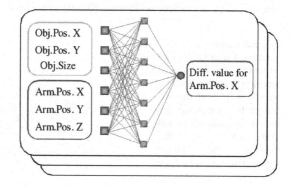

## 5.2.3 Experiments

### 5.2.3.1 Teleoperative Manipulation System

Our manipulation system consists of a 5-DOF manipulator arm (attached to a Pioneer3AT mobile platform), a USB camera, and a laptop for hosting of the robot controller. The Pioneer Arm is a relatively low-cost robot arm that is driven by six open-loop servo motors, providing 5 degrees-of-freedom with an end-effector capable of grasping objects up to 150 grams in weight. For acquisition of the visual data, we mount a small USB webcam on the gripper so our system can transmit the workspace view observed by the end-effector to the user. The maximum frame rate of the camera is approximately 30 fps with pixel resolution of 320x240. It also has a diagonal 54 degrees of field-of-view angle with focus range of 5cm to infinity. The haptic device used by the human operator is a force-feedback joystick (Microsoft SideWinder2 force feedback joystick), which has a 16bit 25MHz onboard processor capable of delivering 100 different forces and 16 programmable function buttons. The haptic device is also coupled with a user interface that receives the streaming images retrieved from the camera attached directly to the robot end-effector.

### 5.2.3.2 Experimental Results

We first validated our force guidance technique by conducting 20 trials for five selected places for object centering (Figure 5.3). The visual data from the camera system mounted on the robot is analyzed every 33ms and a target object is acquired and tracked. The guidance force is then generated directly from the visual data to guide the human operator using the haptic control device to move the arm toward the top of the object. In these trials, the approach ratio is chosen as 0.5 and the dimension of the object is (25mm, 30mm) as measured in the x-y plane.

**Fig. 5.3** Camera view of before and after Object Centering task. The arm is manipulated by a human operator who 'sees' through this view and 'feels' the guidance force directly generated from this visual data. The '+' mark indicates the size of the object, and the box indicates the center area.

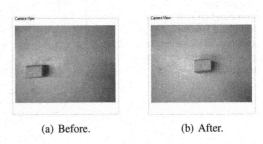

(a) Before.                    (b) After.

Figure 5.4 shows typical graphs of distance change in object centering when the force guidance is enabled or disabled. With the guidance force, we can see the arm is moving toward the object making a clean path, whereas without the force guidance the arm hesitates at the beginning (since the operator has to decide where to move the arm) and takes more time in centering (see Figure 5.5). The average time for object centering (for a certain target position) was 2.1 sec for the first 10 trials and 1.8 sec for the latter 10 when the guidance force was enabled, while it took 3.2 sec and 2.4 sec respectively when the force was off. We can see that the average time with the force guidance is 29% less in total, and 33% smaller in the first 10 trials. The common decrease in the latter 10 trials in both cases shows that the human operator learned to operate better, and we can also see that the average time in the latter part with force guidance was still 25% faster than that without force guidance (see Table 5.1).

We then test our learning approach and show the results derived from applying our learning methodology to an object manipulation task. To train our robotic system to reach down and grasp an object, we acquire training trajectories for 9 different locations of the target object. The number was chosen to provide spatial training data sets, which includes a center position and 8 positions with different directions having distances of 5-7cm from the center position. The average execution time for completion of a sequence was 10.2sec, with the trajectories projects onto the XY plane depicted in Figure 5.6. The resulting ANNs could perform 15 successful tasks out of 20 trials over random target positions (different than those used to train the system), resulting in a performance rate of 75%.

**Table 5.1** Average time for object centering.

| Trials | Time Comparison in Object Centering | | |
| --- | --- | --- | --- |
| | With Force Guidance | Without Force Guidance | Effectiveness |
| First 10 trials | 2124 ms | 3165 ms | 33% Faster |
| Last 10 trials | 1793 ms | 2383 ms | 25% Faster |
| Total Average | 1956 ms | 2774 ms | 29% Faster |

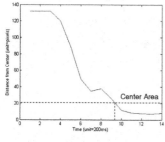

(a) Force guidance enabled.

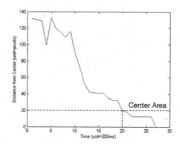

(b) Force guidance disabled.

**Fig. 5.4** Distances from object center in time domain.

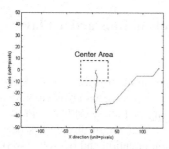

(a) Force guidance enabled.

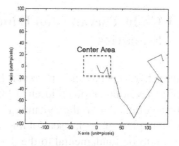

(b) Force guidance disabled.

**Fig. 5.5** Arm movement trajectories towards the center.

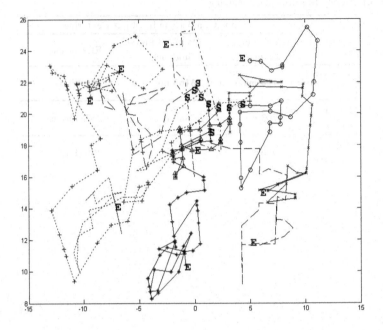

**Fig. 5.6** End-effector trajectories of nine training data sequences. Only XY plane is drawn and Z axis term is omitted here. 'S' is the starting position and 'E' is the final position for each trajectories.

## 5.3 Child Therapy and Education: Robots in Interactive Play Scenarios

Interactive play has an important role in the development of cognitive, physical, and social development in children (128; 75). The effect of playing has shown to have a lasting effect due to the dynamic nature of interacting with the world (76). In fact, the interplay of social scaffolding and self-induced locomotion in play has been shown to be fundamental to the development of joint attention and of 'self' (108). While typically-developing children generally do not need the presence of an adult to engage in play, studies have shown that the interactions of children with other individuals will increase their social development in areas such as sharing and turn taking. By playing with others, shared interest between playmates enables developmental play, engaging the mind, and creating opportunities for extended play over longer durations. As such, developing robots to engage children in play could promote an environment of active learning. Other researchers have shown that robots, as toys, can aid in early intervention for children with development delays (e.g. (52; 149)), assist with physically challenged children (e.g. (99)), and engage children in imitation base play (e.g. (114)). Most robots though, do not physically manipulate toys in conjunction with their human (child) partner. To become an effective playmate, a robot must be capable of fully engaging in manipulative play. This is compatible with the play theory that exploration and manipulation are prerequisites to meaningful play experiences (63).

To develop an effective robot playmate, we begin by developing algorithms to allow the robot to identify general play behavior associated with manipulation tasks related to play activities (87). The basis of our methodology is that if the robot could interpret the basic movements of a humans play, it will be able to interact with many different kinds of toys, in conjunction with its human playmate.

### 5.3.1 Defining Play Primitives

One of the benefits of playing with toys through manipulation is to stimulate the development of fine motor skills, which require control of small, specialized motions using the fingers and hands. These skills evolve over time starting with primitive gestures, such as grabbing (76). By examining the interaction of children with toys, a number of common primitive gestures can be extracted, such as grasping, transporting, inserting, hammering, stacking, and pushing. With regards to a robotic playmate, these primitive gestures are what we further define as play primitives.

In various studies conducted on the dynamics of play, a number of mutually exclusive categories of play have been derived (108; 52). Of these, two are associated with physical manipulation - 1) functional play, which involves the conventional use of objects (Figure 7(a)) and 2) relational play, which is defined as the association of two or more objects together (Figure 7(b)). As such, we define two types of robot play primitives:

- Functional play primitives - hand/fingers manipulate a single toy object.
- Relational play primitives - hand/fingers manipulate a single toy object such that it makes contact with another toy object.

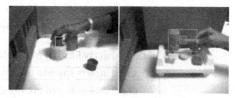

(a) Grabbing.                                    (b) Stacking or hammering .

**Fig. 5.7** Examples of Functional play 7(a), and relational play 7(b).

### 5.3.1.1 Motion Behavior Analysis

To identify individual functional or relational play primitives, we first extract low-level behaviors observed during human playing actions. A motion behavior is used to represent an interpretation of these basic play movements. Motion behaviors are represented by a motion vector, which captures both the *direction* and *velocity* of the motion. The possible values associated with *direction* and *velocity* are discretized based on pre-defined linguistic classes, as depicted in Table 5.2, resulting in a finite set of possible motion vectors (or a motion class). Functional direction represents the absolute direction of the play object with respect to a world coordinate system. Relational direction describes the positive or negative motion of an object relative to a target. Distance between an object and a target is computed when a motion initiates and terminates based on the target identified at the end of a play. These two distances are then compared and define the positive and negative relational direction. The velocity of the motion behavior is measured with respect to the distance between the location of an object when a motion initiates and terminates and the average time associated with the motion. This velocity component is reclassified as a speed: SLOW/FAST, such that if a motion is faster than the overall sequence speed average, it is defined as FAST, and as SLOW otherwise.

**Table 5.2** Motion behavior definition structure.

| Motion Parameter | Play Primitive | Linguistic Values |
|---|---|---|
| Direction (d) | Functional | Left, Right, Up, Down |
|  | Relational | Positive: towards target<br>Negative: away from target |
| Velocity (v) |  | Slow, Fast |

As an illustrative example, Table III shows the association between low-level motion behaviors and the resulting motion vectors.

**Table 5.3** Association between motion behaviors and vectors.

| Illustrative Description of Motion Behavior | Motion Vector |
|---|---|
| Child quickly lifts toy from table | (Up, Fast) |
| Child inserts toy into toy-bin | (Positive, Slow) |
| Child shakes toy to the right | (Right, Fast) |

The motion behavior analysis process involves populating instances of the motion vector based on observation of a human playing action. This process is executed by computing a motion gradient during human play actions and fitting the motion gradient to the pre-defined motion class. Algorithmically, our motion behavior analysis process consists of the following steps: 1) Compute the motion gradient associated with observation of the human. Compute and normalize the motion gradients for N consecutive motion frames to minimize the effect of motion jitter. 2) Find the motion vector which minimizes the least squares estimation of the motion model fit, and 3) Normalize and fit the motion vector to the pre-defined motion class to compute direction and velocity of the motion.

### 5.3.1.2 Behavior Sequencing

In order to extract play primitives for a robot playmate, behavior sequencing involves identifying and labeling the sequence of motion behaviors associated with a play scenario. To perform this sequencing operation, the play toy of interest is first identified by using the highly saturated color parameters of childrens toys to perform steps of RGB to HSV color conversion, histogram back-projection, and segmentation. The play toy, designated as the first one to take an action, is then tracked over subsequent motion frames (Figure 5.8). Using the motion behavior analysis process, the set of individual motion behaviors is then determined and sequenced based on movement of the play object. The final resting destination of the play object is used to identify the final play primitive.

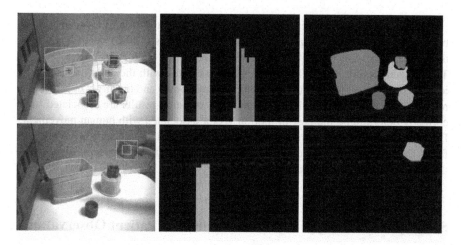

**Fig. 5.8** (Top) Identifying toy objects: Original toy scene, Back-projected hue histogram, Final toy objects detected (Bottom) Tracking the play object: Original toy scene, Back-projected hue histogram based on play object, Final play object detected.

## 5.3.2 Experimental Results

Four human test subjects were asked to perform repeated play primitives for five different play scenarios (Figure 5.9). The human was observed from a fixed camera with a side view. Multiple toy objects were randomly positioned in the play scene and the human was instructed to either 1) select any object and stack onto another (stack primitive) or 2) select any object and insert into another (insert primitive). In three of the play scenarios, the human was instructed to perform these actions continuously (i.e. multiple times) and in two of the play scenarios the human was instructed to perform only one play action. The experiments were designed to test the capability of the system to identify both the motion behaviors as well as the correct play primitives, given the differences in motion behaviors for the different subjects.

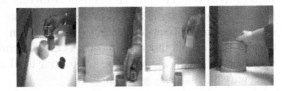

**Fig. 5.9** Some play scenarios.

Given the various test scenarios, the performance of the system was categorized based on correct recognition of play and target objects (100%), identification of the motion behaviors (98.17%), and correct labeling of the play primitive (100%). The play/target object recognition rates are associated with the ability to correctly identify both the play object (upon grabbing a toy object) and the target object (upon releasing the play object). Motion behavior recognition results are associated with correctly determining the best-fit motion vector associated with human movement. Errors in this calculation were primarily due to miscalculation of speed (slow versus fast). The play primitive recognition rate was associated with correctly identifying the play primitive (insert/stack).

## 5.4 Physical Therapy: Robot Assistance via Patient Observation

Activities of daily living (ADL) are predicated upon the successful accomplishment of basic actions or movements, according to the current International Classification of Functioning, Disability and Health (ICF) framework (194; 4). Thus, in this section, we discuss a methodology for monitoring of physical therapy exercises, as a first step in providing feedback for assessing performance improvements in basic ADL (88). Physical therapy, defined by the US Department of Labor as the practice of providing services to help restore function, improve mobility, and limit permanent physical disabilities of patients suffering from injuries, is a very practitioner intensive process. Physical therapy sessions typically transpire over an extended period of time and require patient adherence to the exercises taught to the patient not only during sessions with the therapist but also at home. Patient compliance is strongly correlated with shorter time to recovery as well as reduction of pain in the long term (190). During the time between therapy sessions there are many factors which affect patient compliance, including forgetfulness, lack of motivation, boredom, and lack of instant feedback. To deal with these issues, researchers have shown the positive use of robots in assistive therapy applications ranging from stroke rehabilitation (77) to motor development in children (66).

In many of these applications, if we can correctly identify and match at-home patient exercise behavior based on characteristics learned during the therapist-patient session, we can develop a monitoring mechanism to provide feedback for patient recovery. To enable this capability, we present two methods that utilize image-based observation as a means of gathering sensing information, and classification to identify subsequent patient behavior based on observations during the therapist-patient session.

### 5.4.1 Learning of Exercise Primitives

Similar in nature to the method presented in Section 3 for constructing robot play primitives from motion vectors, we model an exercise scenario by sequencing a series of repetitive motion behaviors together based on observation of a human exercise action. The first stage in the process involves tracking movement of the arm using vision-based marker detection and tracking of a hand-held object (e.g. dumbbell) (Figure 5.10). Markers are detected by color segmentation using methods of Histogram Back-projection and Gaussian Filtering. Among the multiple markers detected, the first one to take an action is considered as the exercise marker. The other markers are then marked as targets, and the motion of the exercise marker is described relatively to these. The motion gradient associated with the exercise marker is computed over consecutive image frames and fit to the set of pre-defined motion classes, as discussed in the previous section, to identify the motion behavior. The sequence of motion behaviors associated with an exercise scenario are stored and labeled (by the therapist). After therapist-patient interaction, the system uses this same process to identify the current exercise scenario and compare with the stored therapy exercise information.

### 5.4.2 Learning of Exercise Behaviors

In the previous approach, image-based methods were used to construct an exercise scenario from a sequence of identified motion behaviors. In the next approach, we utilize a method that classifies the entire exercise scenario using a single rep-

Fig. 5.10 Identifying exercise
and target markers. Blue is
associated with the exercise
marker and green with the
target marker.

resentation. Based on imaging the patient during a therapy session, a texture based
feature vector is first generated for each image (frame) and stored in a database.
This database is then used to train a self organizing map to classify the elements in
the dataset, using the approach as described in (142). During subsequent exercise,
the method presented in (156) is used to extract period and frequency information
for the captured data in order to generate a mapping between observed state and its
position in the exercise cycle. In this step, we assume only one exercise is exhib-
ited in the captured data sequence. After therapist-patient interaction, a measure of
similarity using the 2D Kolomogorov Smirnov test (112) is calculated to determine
the statistical goodness of fit between pairs of exercise behaviors. This test is used
to determine which of the stored therapy exercises the patient is performing during
subsequent exercises.

### 5.4.3 Experimental Results

For testing the performance of the algorithm, video clips of subjects performing
different arm exercises, of varying degree profiles, in seated and standing position,
and with the left and right arm, was collected (Figure 5.11). The three different arm
exercises considered were:

- Shoulder Flexion - Raise arm to point to ceiling, keeping elbows straight
- Shoulder Abduction - Raise arm out to shoulder level, keeping elbow straight
- Shoulder Rotation - Keep elbow in place and slide forearm back and forth

In the first methodology, which identifies exercises by sequencing together motion behaviors, the following test sequence was utilized.

In this test scenario, the algorithm was able to achieve a 100% classification rate,
and was even able to correctly classify test sequence 9 and 10 as incorrect exercises
(due to the patient not achieving the full range of motion).

**Fig. 5.11** Example sequence of images captured during observation (top) 180 degree left shoulder abduction (middle) 90 degree left shoulder abduction (bottom) left shoulder rotation.

**Table 5.4** Test behaviors used for study with first methodology.

| Test | 1 | 2 | 3 | 4 | 5 |
|---|---|---|---|---|---|
| Input Exercise | #1 | #2 | #3 | #3 | #2 |
| Description | 180 Right Shoulder Flexion Seated Position | 90 Right Shoulder Abduction Seated Position | Max Right Shoulder Rotation Seated Position | Max Left Shoulder Rotation Seated Position | 90 Right Shoulder Abduction Seated Position |

| Test | 6 | 7 | 8 | 9 | 10 |
|---|---|---|---|---|---|
| Input Exercise | #1 | #2 | #1 | Incorrect | Incorrect |
| Description | 180 Left Shoulder Flexion Seated Position | 90 Left Shoulder Abduction Seated Position | 180 Right Shoulder Flexion Seated Position | Incorrect Angle Right Shoulder Flexion Seated Position | Incorrect Angle Right Shoulder Flexion Seated Position |

In the second methodology, which labels the entire exercise sequence using a single representation, the following test sequence was utilized.

**Table 5.5** Test behaviors used for study with second methodology.

| Test | 1 | 2 | 3 | 4 | 5 | 6 |
|---|---|---|---|---|---|---|
| Exercise | #1 | #1 | Incorrect | #1 | #1 | #2 |
| Description | 180 Right Shoulder Abduction Seated Position (front view) | 180 Right Shoulder Abduction Seated Position (front view) | 180 Left Shoulder Abduction Seated Position (front view) | 180 Right Shoulder Abduction Standing (left view) | 180 Right Shoulder Abduction Standing (right view) | 90 Right Shoulder Abduction Seated Position (front view) |

| Test | 7 | 8 | 9 | 10 | 11 |
|---|---|---|---|---|---|
| Exercise | #2 | Incorrect | Incorrect | #3 | #3 |
| Description | 90 Right Shoulder Abduction Seated Position (front view) | Incorrect Angle Right Shoulder Abduction Seated Position (front view) | Incorrect Angle Right Shoulder Abduction Seated Position (front view) | Sup. Right Shoulder Rotation Seated Position (front view) | Sup. Right Shoulder Rotation Seated Position (front view) |

The resulting categorization of exercises grouped Exercise 1, 2, 4, and 5 into one class, Exercise 6 and 7 into one class, Exercise 10 and 11 into one class, and 3,8, and 9 each into separate classes. The resulting classification scheme is summarized as follows:

**Table 5.6** Results generated from second methodology.

| Test | 1 | 2 | 3 | 4 | 5 | 6 | 7 | 8 | 9 | 10 | 11 |
|------|---|---|---|---|---|---|---|---|---|----|----|
| Exercise | #1 | #1 | Incorrect | #1 | #1 | #2 | #2 | Incorrect | Incorrect | #3 | #3 |
| Classified w/ | #1 | #1 | None | #1 | #1 | #2 | #2 | None | None | #3 | #3 |

## 5.5 Conclusions

In this chapter, we discuss methods we believe are the first steps for endowing robots with the ability to provide manipulative assistance whether in interactive play or for activities of daily living. As robots continue to hold a greater role in assisting in the rehabilitation, care, and education of others, the necessity for endowing them with improved learning capability poses interesting challenges. Through experimental results, we have shown that there are future possibilities in developing an infrastructure for enabling humans to provide instruction to a robotic mate.

Given the methods discussed in this chapter, one aspect of future work includes integrating interactive learning capability with robot mobility. For interactive play, the structure of the derived play primitive is designed for transmission to a robot platform for subsequent turn-taking in toy manipulation (Figure 5.12). For monitoring in therapy, the robot must have mobility capability to ensure correct image capturing of the patient during execution of each exercise (Figure 5.13). And for teleoperated assistive robots, the range of achievable tasks must include those of a consistent, repetitive nature (Figure 5.14).

**Acknowledgements** This work was supported in part by the National Science Foundation under Award Number IIS-0705130. Seed funding to begin the robotic playmate effort was provided by the Center for Robotics and Intelligent Machines (RIM@GT).

**Fig. 5.12** Robot (Robosapien) platform for interactive play.

**Fig. 5.13** Robot platform for therapy monitoring.

**Fig. 5.14** Experimental setup and tasks for activities of daily living (ADL).

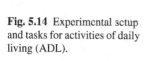  

(a) Food preparation.  (b) Grooming and personal hygiene.

Fig. 5.12 Robot ...
... of applying a ...
... cooking knives.

Fig. 5.13 Robot platform for
... deploy ingredients ...

Fig. 5.14 Experimental setup
and performance of use of three
kitchen utensils: ...

(a) Food preparation.    (b) Covering and personal
                            hygiene.

# Chapter 6
# A New Direction in Human-Robot Interaction: A Lesson from Star Wars?

Gerard Jounghyun Kim

## 6.1 Introduction

Robots are finding their ways into our everyday environment (68) and as such their social aspect and interaction with humans are becoming a very important issue for their successful deployment and control. However, "directly" interacting with robots has thus far been largely unsuccessful.

**Fig. 6.1** Interacting "directly" with robots. Robots may (or perhaps cannot) be ergonomically designed for diverse users. It may also falsely induce expectations of human like responses.

One main reason is that most robots are not (or perhaps cannot be) designed well human factor-wise for diverse users. Figure 6.1 illustrates the case very well. Touch screens (on the robot) are usually small and difficult to reach, and their interfaces are difficult to use. A robot as a moving desktop computer may not be a well thought

Gerard Jounghyun Kim
Digital Experience Laboratory, Korea University, e-mail: gjkim@korea.ac.kr

out metaphor despite our familiarity with the desktop interfaces. On the other hand, today's robots tend to pursue human-like communication means while still lacking much in most such abilities. Robust vision-based or voice/language-based robot communication seems very difficult in practice. The problem of anthropomorphic interfaces (of which robots are usually) generating false expectations of human-like responses and behaviors from robots, and becoming a source of disappointment and confusion to users, has been well known (18; 122).

Figure 6.2 shows shots from the movie Star Wars (3) in which the robot R2D2 uses a 3D holographic projection to convey a message from Prince Leia. Even though the movie is only an imaginary plot, it is quite plausible to suggest that such "non-human-like" communication means can be more effective, at least for certain tasks or situations. Interestingly, in this particular shot, C3PO, a perfectly human-like robot, also listens in to the projected message.

**Fig. 6.2** Shots from the movie Star Wars (3), and the robot R2D2 with a holographic projector. Even in the futuristic days, 3D projection is depicted to be more effective (or dramatic) than using human-like communication (e.g. notice that the humanoid robot, C3PO, also is drawn to the situation).

While robotics research must be continued in the direction to produce human-like capabilities, we propose (inspired by ideas from Star Wars) an alternative, that is, an "indirect" human-robot interaction (HRI) mediated through a hand-held device and a projection display (since 3D hologram technology is still immature). Figure 6.3 shows one possible way to employ such a framework where the robot projects onto a donut shaped area on the floor to serve and interact with multiple users at the same time. Such an HRI method could be useful for large but sparsely populated areas like conventions, airports, post offices, etc. The robot can move around and stop to offer interactive services by using a camera (and/or other sensors) to detect anybody in the vicinity who may be interested to interact. Figure 6.4 shows another possibility in which a robot actively guides a group of people in a museum. The robot, given the map of the museum and a predetermined path, with access to localization information for itself and the people surrounding it (using onboard sensors or infrastructure), can select an appropriate surface to project information for the users and at the same time provide a medium through which users can interact with the robot (e.g. using a laser pointer or a cell phone as an interaction device).

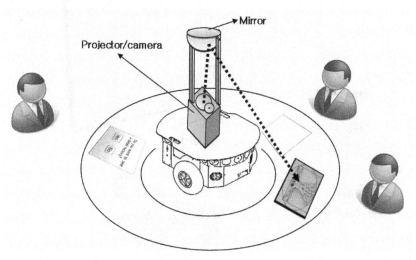

**Fig. 6.3** A robot projecting to a donut shaped area to serve multiple users. The robot may roam around and stop to serve by detecting (using sensors/camera) any potentially interested users.

We claim that such an interaction framework can alleviate much of the human factors problem and enhance the overall experience by the "enlarged" communication channel and amplified presence of the robot. In addition, the proposed framework is flexible to accommodate different types of users and tasks, one-to-many interaction, even mixing of real and virtual objects, and when combined with a hand-held interaction device, it can be extended to support a variety of interaction techniques with the consideration of robot's mobility. In this paper we concretize our proposal and discuss various important technical issues in realizing it.

## 6.2 Indirect Human-Robot Interaction

As already indicated, our framework suggests to use the projection display as the mediating element between the human user and the robot. Figure 6.5 illustrates the indirect HRI framework. There are several technical challenges for this proposed framework in each of the respective constituent parts.

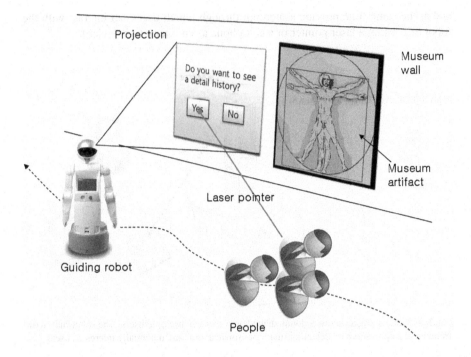

**Fig. 6.4** A robot guiding a group of people in a museum. Depending on the robot's position, floor and object layout, and observer's position, the robot can appropriately select the surface to project useful information and at the same time provide a physical medium for interaction to occur using a simple hand-held device like a laser pointer or a cell phone.

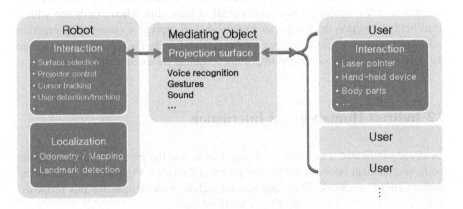

**Fig. 6.5** Major technical challenges of the proposed indirect HRI framework: (1) robot tracking, (2) user/environment sensing (3) flexible display projection, and (4) projection surface centered multimodal interaction method.

## 6.2.1 Robot location/pose tracking

First, the robot must be able to track itself in relation to at least its immediate environment. This is because the robot must determine an appropriate surface in the environment to project onto. A technique called the "Simultaneous Localization and Mapping (SLAM)" is a popular method to solve this problem (176). SLAM is based on augmenting the dead reckoning (e.g. from odometry of the robot) with location prediction through landmark detection and probabilistically building the environment map at the same time. Assuming a map of the environment exists, even more accurate location tracking is possible. Other tracking or localization techniques are certainly applicable. While theoretically it is possible to build the environment map and extract surface information, we can often safely assume that a 2D map of the robot operating area is given.

## 6.2.2 User/environment sensing

Secondly, the robot must be able to sense various aspects of the users from a nominal distance (e.g. 0.5m~3m). For instance, the user's position in relation to the robot is important to determine the appropriate place to project to. It is relatively easy for a robot to recognize the existence and approximate relative location of nearby users (or obstacles) using ultrasonic sensors or cameras. In addition, the viewing direction of the user, user's motion characteristics, user's identity (e.g. through face recognition) may all be used to tailor and customize the display/interface contents. Equally important is the cursor tracking for interaction. Since the proposed mediating interaction element is the display surface, the most natural way to interact is through a 2D cursor. Simple methods of camera based tracking of laser pointers, special markers, body parts and object occlusion can be employed with relative ease (188; 101). Note also that landmark detection is also needed for accurate robot localization as already mentioned.

## 6.2.3 Flexible projection

The third important problem is making the projection. First, the robot must select the appropriate surface in the environment in a stable and convenient manner (for the user) as it moves along. Such a selection will depend on the changing absolute and relative positions of the robot and the users. In addition, since the robot is moving, the projection display is also moving. One must consider aiding the user in some way to minimize any negative effects due to this characteristic (e.g. render a small arrow explicitly indicating and drawing the attention of the user to the movement of the display). Once the projection surface is selected, a control mechanism (e.g. pan and tilt) would be needed to position and aim the projector toward that surface.

Most commonly the projection axis and the target surface will not be perpendicular, thus an off axis projection must be made. In fact, the surface may not even be flat or could be angled. A related issue is rendering the visual content with the detected user viewpoint in mind. Fortunately, many algorithms have been devised to handle such non-planar surface projection (131; 139; 178).

### 6.2.4 Large display surface centered interaction design

Although the large display surface serves as a place at which standard cursor based interaction can occur, the overall framework is quite different from the desktop computing environment. For example, in Figure 2, the robot may be serving multiple users and it is not immediately clear how sound feedback can be employed at the same time (e.g. whether it should be used at all, or if directional sound be discerned by the respective users). When projecting on to the floor, it can be effective to render the contents with a 3D look as if they are standing on the floor for better visibility. In the museum guidance situation of Figure 6.4, the robot may be moving as it tries to convey information and interact with people. The right interplay between moving robot and users would be one of the keys to producing a usable interface, one of the original goal of this work.

## 6.3 Summary and Postscript

We have proposed a new indirect HRI framework based on the idea of mediating the interaction through projection from the robot. Our work is motivated by the lack of usability with the current mostly direct approach to interaction design for HRI. We have streamlined major technical issues, and believe that the proposed framework can serve as a good alternative to the current form of HRI practice.

I believe that it is not coincidental that I have developed this new interest in human-robot interaction. Back when I entered the USC Robotics Laboratory to become Professor Bekey's student, my research interests were more in the traditional aspects of robotics. But, one thing led to another and I wound up investigating design and manufacturing problems as related to robotic assembly. This later took me into the field of virtual reality and simulation, and then 3D multimodal interaction. Thus, it seems only natural that I have traveled full circle back to the field of my heart. Looking back, I truly believe that my studentship under Professor Bekey and association with the outstanding bunch of colleagues at the USC Robotics Lab still form the backbone of my capabilities as a research professor and educator, particularly in terms of developing a wide perspective, knowledge basis and strong initiatives (which allowed me to successfully venture into many interesting and related fields).

# Chapter 7
# Neurorobotics Primer

M. Anthony Lewis and Theresa J. Klein

**Abstract** Neurorobots use accurate biological models of neurons to control the behavior of biologically inspired or biorobots. While highly simplified neural models (e.g. ANN) have been used in robotics, recent innovations in mathematics, science and technology have made possible the real-time control of robotics by highly complex, biologically realistic neural models. In this chapter we present a self-contained primer on Neurorobotics which serves to give an integrated view of the possibilities of this nascent trend with important ramification in science and technology. In particular, we argue that neurorobotics will replace the conventional computer simulation for many neural-system models. Further, within a relatively short-time it will be possible to simulate $10^{11}$ neurons in real-time, roughly the number of neurons in the human brain, on a desktop computer. If we can understand how to harness this power, and productize it, we will be able to create robots of incredible complexity and behavioral richness.

## 7.1 Introduction

I can not believe that the brain computes Jacobians - George A. Bekey circa 1992

Biorobots are artificial devices built to test hypotheses in biology. Examples include work by Webb (193), Beer (25; 24) Lewis (107; 106; 105) and many others. Biologically inspired robots, on the other hand, are robots that use biology as metaphors to solve practical problems in robotics. Examples include work by Brooks (35; 36)

M. Anthony Lewis
Department of Electrical and Computer Engineering, University of Arizona, Tucson, AZ, 85721, e-mail: malewis@ece.arizona.edu

Theresa J. Klein
Department of Electrical and Computer Engineering, University of Arizona, Tucson, AZ, 85721, e-mail: tjk@ece.arizona.edu

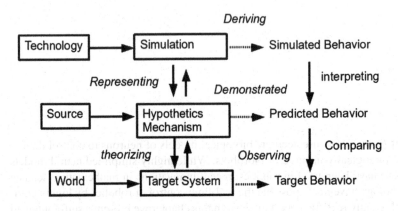

**Fig. 7.1** Integration of simulation studies with the scientific processes. Redrawn from (193).

and Arkin (16). A neurorobot is a special kind of biorobot that explicitly uses models of biologically realistic neurons as its computational substrate. In a neurorobot, algorithms are replaced by extremely high dimensional dynamical systems based on neural sub-units. These subunits range in complexity from leaky-integrator models to integrate-and-fire models, Hodgkin-Huxley neurons and the recently discovered model by Izhikevich (90). Neurorobots will one day solve real-world problems, thus filling a dual use as both a biorobot and an biologically inspired robot.

Neurorobots, an outgrowth of biorobots, may find important commercial applications in biologically inspired robots in the near future. For a comprehensive overview of biologically inspired robots, see (27).

## 7.1.1 Neurorobots and the Scientific Method

Can neurorobotics help us better understand the brain? Webb (193) has proposed a model for integrating the traditional view of the scientific method with modern technology, in particular, simulation. Referring to Fig. 7.1 the process of scientific investigation begins with the identification of a target system that we are interested in understanding. The scientist may theorize a mechanism that will explain the observed target behavior. The scientist may use various sources for inspiration for this mechanism. In biomechanics, it may be a cost function that we presume the human body optimizes. In the example cited by Webb, the idea of a Fourier transformation may be used as a source for a hypothetical mechanism of how the cochlea transforms sound into neural impulses.

To understand the ramification of this hypothesis, we can create a simulation that takes similar input and gives similar behavioral output. A simulation can exist solely in a computer, or, in the case of a biorobot, the system can interact with the world. As computers are, of course, capable of fantastically faster mathematical computations than the human brain, one may think of a simulation as cognitive prosthesis that helps us think more effectively about complex problems that are reducible to a well defined system of equations.

A biorobot is a simulation paradigm for understanding behavior in animals in which a computer program is allowed to gather and process raw data from the world in real-time and produce an effect on the world. A biorobotic simulation is as legitimate a scientific tool as a self-contained computer simulation. Ultimately, the difference between a simulation that runs in a simulated world and a simulation that runs in the real-world is that we can compare our physical simulations with the target system in the same environment as the organism we are investigating. *For this reason, we might say that for understanding the neural basis of behavior, a biorobot can be a more legitimate and scientifically meaningful simulation, than a computer simulation.*

## *7.1.2 21st Century Robotics: Productizing Mythology*

The idea of a robot is grounded in biological inspiration. It is a search for what constitutes the "spark of life" and what distinguishes the living from the non-living. It raises the question, "how can we create artificial beings?" Automata have been built for thousands of years to amaze and entertain and make promises of slave-like machines that will serve human-kind. It was only in the 20th century, with the advent of the stored program, digital computers that we could begin to realize practical, commercially viable robotic machines. Stored program digital computers enable the following capabilities in robots:

1. *The ability to create highly complex behavior-* Complex behavior is dependent on the ability to make memories. In the case of computer programs, memories can be constructed from bistable elements integrated on a massive scale onto inexpensive chips. Complex behaviors uses a large number of non-linear processes to make decisions, compute transformation and most importantly to derive percepts. A percept is an element extracted from a sensory stream that captures some invariant property of that sensory stream (104). We distinguish this from a linear transformation of an input stream where the original sensory input can be recovered from the transformed elements. Many different sensory configurations can create the same percept. This processing is easily and reliably simulated on digital computer.
2. *The ability to alter behavior to achieve different tasks (i.e. to be reprogrammed)-* A stored program computer can be altered in its operation by controlling the

data it operates on, controlling the mode of the code that is executed and so on. These are properties that are exceedingly difficult to achieve without modern computers.

To date, the most aggressive uses of the ability to perceive have been seen in autonomous robots and factory automation. However, at the level of consumer robots, there has been marginal ability to perceive the environment, little ability to be reprogrammed, and relatively simple behavior. Thus, the consumer robots have not fully utilized the power of the modern computer. Why?

Productization of highly complex machines is highly problematic. In general, the *Occam's razor principle of productization is that the simplest device is the one that will most likely be built.* Thus, introducing complexity is contrary to the parsimony of productization.

From a product perspective the industrialist must answer several question:

1. What advantage does complexity give us? Does it make the product better or worse?
2. How does one test such a machine to know if it is working?
3. How can we tell if a learning system is learning?
4. If a system requires thousands of cycles to significantly change behavior, how can we test such a capability quickly?

The problem will be complicated even further in the case of a neurorobot. At the end of the factory line, tests must be done to confirm the behavior of the robot being built. Suppose that a robot system had the complexity of a human being. Much as the human mind is 'programmed' by a sequence of experiences, neurally based robots will be programmed as well by experience. Their resulting behavior will be difficult to predict. It might take 16 years to determine if the robot would be a juvenile robotic delinquent or is on track to be a brain surgeon! Debugging, at the system level, will require new automated tools that can analyze the system using invasive techniques, or perhaps an entire field or robot psychology.

For these reason neurorobotic technology will be very limited in market acceptation, or indeed other highly intelligent machines, until we can solve the practical problem of testing such systems, and demonstrating that these systems will yield clearly superior performance— performance that can be achieved in no other way more simply. One exception to the rule on market acceptance has been the robots built by Mark Tilden. Tilden's first prototype of the 'Robosapien' line of robots was built at the Neuromorphic engineering workshop in Telluride, Co, in July, 2001, see Fig. 7.2. Some 5-10 million Robosapien robots and its derivatives have been sold. The essence of Tilden's idea was to use analog oscillators to generate the movement of bodies. His robots, while remarkably sophisticated as a consumer product also exhibit a *je ne sais quoi* quality of life-like movement that has not been duplicated by others. His prolific experimentation in the 1990's lead to the development of the first prototype Robosapien in July of 2001.

Tilden's robots were acceptable as products but had only limited perceptual capability due to the sever cost constraints of the toy market (not, indeed, by any limitation of Tilden's imagination).

**Fig. 7.2** The birth of a revolution in biologically inspired robots. (A) The first Robosapien prototype was created in July 2001 in Telluride Colorado by Dr. Mark Tilden. Tilden successfully incorporated many principles of biological motion into the a commercially viable product. (B) November, 2005 People's Republic of China, the second version of RoboSapien, RS-V2 is being assembled, tested, and readied for shipment. Between 5-10 million robots, based on this design, have been sold (198).

To date, Tilden's commercial efforts have been limited to the consumer entertainment market. It is evident that neurorobotics greatest impact will be both in science and in the service robotics market, or robots that perform useful work.

Productization of a myth is exceedingly difficult, but progress is being made.

## 7.1.3 Computational Substrate

Neurorobots, by definition, must function in real-time. As a result, computational speed and I/O bandwidth constrain the level of complexity now achievable in neurorobotic brains. The substrate for computation is important. Computation used in Neurorobot may be based on computer simulation or may be directly implemented as Neuromorphic chips.

## 7.1.4 Neuromorphic Chips

A neuromorphic chip is an analog VLSI device that implements a specific neural-system computation or class of computations. Historically, these neuromorphic chips differ from their digital cousins in two essential ways. First, these systems rely on analog computations that directly implement operations such as integration, differentiation, multiplication and addition with the purpose of efficiently implementing neuronal networks. Second, digital computers that we are familiar with are stored program computers. They are reconfigurable and capable of running an extremely wide range of computations. Neuromorphic devices have lacked this capability. Neuromorphic chips trade speed and power efficiency for flexibility and development ease. Each chip may have only a handful of people in the world that understand how to use a given device. This limits their dissemination.

### 7.1.5 Graphics Processing Units

Another approach to computation in Neurorobots is to use conventional Central Processing Units (CPU) and Graphics Processing Units (GPUs) for processing. GPUs are leading a revolution in computing. GPU chip concentrate most of their real-estate on computation rather than memory as in traditional CPUs. This allows 16 computational cores to be placed on a single chip. What is more, these chips are inexpensive. The retail price for a state of the art Pentium CPU with 4 cores is roughly the same as a GPU card with some 240 cores. GPU processing have doubled in speed consistently every year (2). Today, a GPU is the most important computational element in a high performance computer, well overshadowing the raw computational power of the CPU. As stated, GPUs are optimized for computation, not memory. This make GPUs an ideal host for complex biophysical algorithms. GPUs will allow the use of models with dynamic synapses, multiple compartments and other advanced features that, to date, have not been incorporated into neurorobotics due to real-time requirements. This revolution in GPU technology will have a major impact in neurorobotics, where computational dense models need to be constructed.

### 7.1.6 Purpose of this Chapter

It is clear that neurorobotics is on the cusp of a revolution due to the aforementioned technological advances. The dissemination of neurorobotic experimentation is limited due to the relatively small group of individuals in the world that are cross trained in robotics and neurocomputation. This article is meant to be a primer in neurorobotics, bringing together fields of robotics, neurocomputation, and computational technology.

It is the hope that this chapter provides the first integrated view of Neurorobotics. This chapter is based on a lecture given at the 2008 Telluride Neuromorphic Engineering Workshop sponsored by NSF.

## 7.2 Classical Robotics

We define classical robotics as the core topics featured in text books by Craig (47), Paul (126), or Asada and Slotine (17). The classic topics consists of (1) configuration space (2) kinematics and inverse kinematics (3) Differential motion (4) Statics (5) Dynamics and (6) trajectory generation. In recent years Baysian probability has come to the forefront as well, and might be considered as apart of the core topics in robotics.

### 7.2.1 Configuration Space

The starting point of classical robotics is configuration space. A robot is an articulate system with numerous rotational and prismatic joints. If the position of each joint is fixed, the system will be in a particular configuration. The set of all possible joint positions of the robot is the configuration space. Picking a point in configuration space defines the position of each point on the robot. You might consider a luxo lamp. By fixing the rotation of the base, the rotation of the light, and the joints in between, the robot has a well defined position or configuration. The configuration space point also sets the position of every point on the surface and in the interior of the robot. The former being important when considering collisions with the environment, and the latter when considering the mass property of the robotics system.

Often, we wish to know how each of these points correspond to a world frame of references, i.e. in Cartesian space. Cartesian space is a simple, generic way of specifying the location of any point on the robot. Cartesian space considers 3 mutually orthogonal axes that define a coordinate system in space. You might think of the corner of a room where the vertical seam is the z axis, and the horizontal wall-floor seams are the x and y axes. Kinematics can help us determine, systematically, the relationship between configuration space and Cartesian space.

### 7.2.2 Kinematics

Refer to the leg in Fig. 7.3. This leg is composed of two segments of length $L_1$ and $L_2$. The position of the foot relative to the hip can be computed if we also know the angle between the axis of each segment and the ground. These angles are given by $\theta_1$ and $\theta_2$.

Thus the hip's position can be computed as:

$$x_e = L_1 \cos(\theta_1) + L_2 \cos(\theta_2) \tag{7.1}$$

$$y_e = L_1 \sin(\theta_1) + L_2 \sin(\theta_2) \tag{7.2}$$

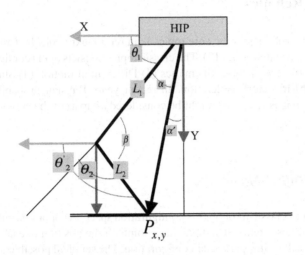

**Fig. 7.3** Two degree of freedom leg. We consider the application of basic techniques in classical robotics to a simple, one legged robot.

The only subtly here is that we have purposefully placed joint angles $\theta_1$ and $\theta_2$ in the frame of reference of the foot. This makes the addition particularly easy. In the typical case, however, the robot may have a sensor on each joint indicating the angle of the one link relative to another link. For the current, planar case, we can make the substitution:

$$\theta_2 = \theta_1 + \theta_2 \tag{7.3}$$

where $\theta_2$ is the angle of the second link as would be measured by a sensor on the robot. We make the simplifying assumption that the hip is always level.

Cartesian space has three axes and can be augmented with a rotation to create the pose of a portion of the robot. For any point in configuration space, there is a unique point in Cartesian space. Further, this map is smooth at almost every point. By systematic analysis of the geometry of a robot, we can compute this map. This is the robotics kinematics.

The magic of kinematics is that if we specify the configuration of our robot using standard notation and apply a standard set of rules, we can turn the metaphorical crank and determine solutions to forward kinematics, that is the mapping from configuration space to world coordinates.

The main point here is that *forward kinematics can be found by well defined procedures.* Inverse kinematics, or going from Cartesian coordinates to configuration space requires some insight. Referring again to Fig. 7.3, we can image that we are given a vector $P_{x,y}$, specifying were we would like to place the foot relative to the hip. The length of this vector is just: $p = \sqrt{x^2 + y^2}$. We now have a triangle with three sides of known length. Applying the law of cosines we have:

$$\alpha = \cos^{-1}\sqrt{\frac{L_1^2 + p^2 - L_2^2}{2L_1 p}} \tag{7.4}$$

$$\beta = \cos^{-1}\sqrt{\frac{L_1^2 + L_2^2 - p^2}{2L_1 L_2}} \tag{7.5}$$

$$\alpha = \arctan 2(y, x) \tag{7.6}$$

$$\theta_1 = \frac{\pi}{2} - \alpha - \alpha \tag{7.7}$$

$$\theta_2 = \pi - \beta \tag{7.8}$$

Thus we find that we can define the angles. In general, as more links are added the inverse kinematics solution becomes exceedingly difficult and requires an increasing level of skill to find a solution.

## 7.2.3 Differential Motion

The mapping from configuration space to Cartesian space is continuous and it is also differentiable. Thus may wish to calculate:

$$dx = f(d\theta) \tag{7.9}$$

This relates small changes in $\theta$ to small changes in $x$. The two are related by a *Jacobian Matrix*:

$$\begin{bmatrix} dx \\ dy \end{bmatrix} = \begin{bmatrix} \frac{df_x}{d\theta_1} & \frac{df_x}{d\theta_2} \\ \frac{df_y}{d\theta_1} & \frac{df_y}{d\theta_2} \end{bmatrix} \begin{bmatrix} d\theta_1 \\ d\theta_2 \end{bmatrix} \tag{7.10}$$

This can be summarised as :

$$dx = J(\theta)d\theta \tag{7.11}$$

Using equations (7.1) and (7.2), we can compute as:

$$\begin{bmatrix} dx \\ dy \end{bmatrix} = \begin{bmatrix} (L_1\sin(\theta_1) + L_2\sin(\theta_1+\theta_2)) & (L_2\sin(\theta_1+\theta_2)) \\ (L_1\cos(\theta_1) + L_2\cos(\theta_1+\theta_2)) & (L_2\cos(\theta_1+\theta_2)) \end{bmatrix} \begin{bmatrix} d\theta_1 \\ d\theta_2 \end{bmatrix} \tag{7.12}$$

What is the relationship between small changes in Cartesian coordinates and small changes in configuration space? It is simple:

$$J(\theta)^{-1} dx = d\theta \tag{7.13}$$

Well, it is simple if it exists. We can find the Jacobian inverse as long as the determinant of $J(\theta)$ is nonzero, i.e. $det\, J(\theta) = 0$. That is, we are checking for the condition:

$$0 = L_1 L_2 (\sin(\theta_1 + \theta_2)\cos(\theta_1) \quad \cos(\theta_1 + \theta_2)\sin(\theta_1)) \tag{7.14}$$

This happens when the robot's leg is held straight out, i.e.: $\theta_2 = 0$ and :

$$0 = L_1 L_2 (\sin(\theta_1)\cos(\theta_1) \quad \cos(\theta_1)\sin(\theta_1)) \tag{7.15}$$

which is always true. Intuitively, when the leg is stretched out, we cannot make a small movement, $d\theta_1$ or $d\theta_2$ that will result in the end of the leg moving away from the hip radially.

### 7.2.4 Statics

Suppose we wish to find a relationship between the force produced at the foot of the robot, and the torques exerted around the knees and hip.
    This relationship is given as:

$$J^T(\theta)\mathbf{F} = \tau \tag{7.16}$$

    where $F$ is a vector of forces at the robot foot, in our example, and $\tau$ is a vector of torques placed at the knee and hip. We omit the proof here, but the reader is referred to (17). This equation might be useful for determining the expected torque on actuators given that the leg is in different configurations. Torque, $\tau$ is just:

$$\tau = \mathbf{r} \quad \mathbf{F} \tag{7.17}$$

    Of course we can find the inverse relationship as well:

$$F = (J^T)^{-1}(\theta)\tau \tag{7.18}$$

    As noted above, the Jacobian will not have an inverse when the leg is straight out. The reader can verify that in this case we will not be able find a torque that will cause the leg to produce a force straight out from the hip. Intuitively, the reader knows this is true. Try standing on your heels with legs fully extended, and then try to jump! Your heel is supporting your weight, but that weight does not cause torques on hips or knees in this example.

### 7.2.5 Redundancy

The most often case in biologically systems is that there are a large number of ways of reaching the same point. As the reader can readily verify, for the simple case of two links, there are two possible configuration for knee and hip for the foot to reach a given point (even if one might represent a painful hyperextension of the

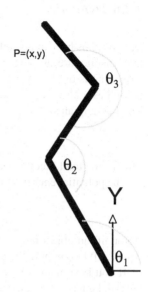

**Fig. 7.4** Illustration of redundancy. In biological systems, redundancy is the norm. Anchor your feet and grab a door handle. You should be able to move your wrist, elbows, torsos in many different ways while keeping your feet anchored and still holding the door handle. Make sure no one else is around when you try this experiment.

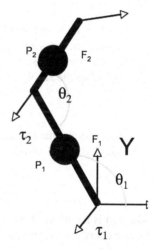

**Fig. 7.5** Model of dynamic system. Here we assume that the mass are concentrated as point masses (illustrated as spheres). Note that when computing Jacobian, we are interested in the distance to and velocity of these masses, not the endpoint of the robot.

knee). Figure 7.4 gives an example of a simple, redundant system where infinite configuration correspond to a given target point.

Humans take advantage of redundancy when they control they posture during movement. Depending on the task, a person might change the configuration of his/her body to be able to exert more force, or remain balanced etc.

## 7.2.6 Dynamics

Real systems have mass and evolve through time, that is, they have *dynamics*. While the dynamics are simple, only in the case of a 1 or perhaps 2 link system, the methodology that is used to write down closed form dynamics is common to both a 1 link as well as a Nth order system. Referring to Fig. 7.5, first we form the *Lagrangian*. The Lagrangian summarized the dynamics of a system.

$$L = KE^{TOTAL} \quad PE^{TOTAL} \tag{7.19}$$

That is, the total kinetic energy minus the potential energy.
    The potential energy of a particle is:

$$PE = mgh \tag{7.20}$$

If you lift an object in a constant gravitational field, with magnitude $g$, one can use eqn (7.20) to compute the potential energy of this object.
    If you have many objects, or in our case, more than one link, you can sum the potential energy. For a mass, we consider the height of the center of gravity (mass). The total potential energy is thus:

$$PE = g \sum_i height_i m_i \tag{7.21}$$

We may with to use some of the machinary developed above to help us compute the height of the mass above the ground, which we can do using forward kinematics, and the velocity of these point masses, which we can do by knowing the joint velocities and using the Jacobian to find the point mass velocities. Eqn. (7.11) implies:

$$\dot{x} = J(\theta)\dot{\theta} \tag{7.22}$$

Let us assume that the mass of each link of the leg can be considered as a single, point mass, then the kinetic energy is:

$$KE = \frac{1}{2}mv^2 \tag{7.23}$$

where m is the mass of the particle, and v is its velocity. One way of looking at the kinetic energy is that it summarizes the net work done on the particle up until that point.
    The kinetic energy of the leg is given as:

$$KE = \frac{1}{2}\dot{x}^T M \dot{x} \tag{7.24}$$

which, using eqn (7.22) we can re-write as:

$$KE = \frac{1}{2}\dot{\theta}^T J(\theta)^T M J(\theta)\dot{\theta} \tag{7.25}$$

here $M$ is a matrix with diagonal elements equal to the mass of each link. In our case, $M$ is to dimensional.

We can now compute the dynamics using a formula:

$$\tau = \frac{d}{dt}\frac{\partial L}{\partial \dot{q}} \quad \frac{\partial L}{\partial q} \tag{7.26}$$

This formula relates the torques at each joint to accelerations, and gravity's pull on each link. At low velocities, gravity is the dominate contributor to torque, far outweighing any other dynamic effects.

## 7.2.7 Trajectory Generation

We must move the limb from point A to point B. If we suddenly accelerate the limb as quickly as possible to the goal position, we will run into problems. First, the exact trajectory of the limb will be ill-defined. We may collide with the environment, or the robot may even colide with itself. Second, we induce a lot of wear and tear on the robot's joints and gears. It is better to choose a smooth trajectory that will allow us to avoid obstacles, and gently accelerate.

The basic idea is to generate a spline from a series of knot points. See Fig. 7.6. The spline has additional constraints placed on it including the position, velocity and acceleration of beginning and end points. The knot points can be positions to avoid obstacles. In classical robotics, the robot control system ensures that trajectories are tracked well.

Note that in classical robotics, we were concerned with controlling the movement of a manipulator. For tasks such as welding is critical that we can achieve good trajectory tracking.

To do this, certain design features are incorporated into the robots. First, we might use high gear reduction motors, perhaps 1:100 or more. This means the robot's joint might revolve 1 times per 100 revolution of the motor. High gear reduction has certain benefits. These benefits include good disturbance rejection: dynamic force perturbations are reduced dramatically as they are reflected back to the motor. Second, we can run the robot at high RPM. In general, electric motor achieve their greatest power output at high RPM. Finally, we would include a stiff feedback control loop. Feedback is the process of comparing the actual position of the a joint to the desired position of the joint and adding or subtracting torque to achieve a reduced error in the next control cycle. In general, disturbances of the environment are repressed.

In a walking machine, if a leg strikes an obstacle while it is in swing phase, two thing can happen: (1) The leg moves out of the way, around the obstacle or (2) the induced torque caused by the strike destabilizes the robot and causes it to stumble.

Nature uses two methods to keep from stumbling. First, in swing phase the leg is very complaint. That is, its feedback gain is very low. Thus, when it strikes, the strike causes a motion of the leg, not the entire robot. Second, animals have reflexes. These

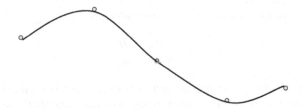

**Fig. 7.6** Trajectory Generation. Trajectories can be specified by a series of 'knot points.' Trajectory generation is the process of finding a smooth path, a spline, through these knot points, while satisfying some endpoint criteria such as position, velocity and acceleration.

reflexes cause the leg to move back, up and over the obstacle. This is an automated response.

In classical robotics we do not consider reflexes except in the case of careful, slow movements,i.e. guarded moves. We do not alter the stiffness of the control loop during trajectory execution.

The next thing to notice about the classical approach to trajectory generation is that knot points are selected by task criteria and not by energy criteria. We have shown that a leg can generate its own trajectory by allowing passive movement of the leg during swing phase (106). Thus the inherent dynamics of the system selects the best trajectory and takes a minimum energy path.

The final thing to notice is that trajectories generation by knot point do not explicitly take into account the cyclic nature of locomotion. Locomotion is highly periodic. We touch on Central Pattern Generator Theory below as an alternative to the classic robot knot-point paradigm.

### 7.2.8 A Pause to Reflect

The power of analytical methods are illustrated here by the ability to determine the overall endpoint of the robot given configuration of subcomponents, i.e. the links. We take can take advantage of some simple properties of vector spaces to allow the addition of subcomponents to find the overall forward kinematics, for example. For inverse kinematic mapping we cannot do this and it is therefore a more difficult in practice.

We note that the Jacobian is a particularly handy transformation. It allows us to tell when the system is capable of generating force at the foot by applying torques at the ankle. It can tell us if the arm has reached a singularity and is therefore not capable of motion in a certain direction.

We note that nowhere in the classical method did we explicitly represent time and synchronization between robot and the environment and the robot and itself.

This mapping problem is non-trivial when we consider tasks such as interaction of any point on the skin of the robot and the world. Here, the kinematic-inverse kinematic computations are so numerous that they cannot be calculated by hand. We must resort to massive computer computations. We note that much of what we uncovered are ways of creating mappings from one space to another. This can easily be learned using even traditional neural networks.

While the purest may argue that the geometric approach is sufficient, its appeal wanes as the complexity of problems grown. Learning methods, including those based on neural building blocks, will become indispensable tools in the roboticists toolbox. CPG methods offer an approach to the time synchronization problem. As CPGs become better understood, they to will become important, indispensable tools.

Artificial Neural Networks have been shown to be universal function approximators. (85). The Universal function approximator is a powerful property that can allow us to subsume much of conventional, geometric based robotics. In the next section we discuss more complex models that allow us to go beyond just simple mapping to create system of reach temporal behavior.

## 7.3 Basic Neurocomputation

Cells are the basic building block of multi-cellular organisms. Special cells exist that are electrically excitable. These cells include muscles cells and in neural cells (Neurons).

The cell membrane (Fig. 7.7) can be thought of as a bag containing the cellular machinery that allows an imbalance in ion concentration to occur inside versus outside the cells. Computationally we do not care about the cells DNA, mitochondria, cell nuclei etc. We do care about this ion imbalance. The cell membrane itself has a very high resistance. On either side of the cell membrane is a conductive solution, thus the cell membrane forms the dielectric of a capacitor. Ion channels penetrate the cell bilipid (composed of two layers of fat molecules) membrane layer and are highly selective to specific ion species.

### 7.3.1 Information Flows into Dendrites and Out of Axons

Neural cells have two types of processes that are attached to the cell body. The *axon* is a long process that can transmit *action potentials*, discussed below, from the cell body to other neurons throughout the brain. These axons synapse on the dendrites of other neurons. These dendrites collect information from thousands of neurons. Each synapse contributes a bit of current to the cell and results in the alteration of the cell membrane voltage. Each synapse may have a different coupling strength. A given cell can either make it more likely or less likely that a cell will generate an action potential, fire, depending on if it is an excitatory synapse or an inhibitory synapse.

## Extracellular Space

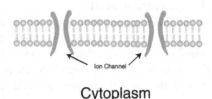

## Cytoplasm

**Fig. 7.7** The Cell membrane is composed of a bilipid insulating layer and isolate conductive electrolytes inside and outside the cell. Certain ion species, such as potassiam, calcium and sodium are particularly important to electrical activity in neurons. The cell membrane contains ion channels that are highly selective to specific ions and can be turned on and off by changes in voltage or by a special protein (ligand). The cell membrane also contains ion pumps that maintain a specific ion imbalance between the inside and outside of the cells. Opening ion channels take advantage of this imbalance to create current flows and hence alter membrane potential. Highly dynamic events called action potential encode voltage events and propagate this information to thousands of other cells.

### 7.3.2 The Neuron Cell is a Capacitor with a Decision Making Capability

The membrane/electrolyte system forms a capacitor. The electrical model of an ideal capacitor is:

$$C\frac{dV}{dt} = i \tag{7.27}$$

where $C$ is the capacitance, $i$ is a current flow, and $V$ is the voltage potential measured relative to the inside of the neuron.

Ions flow is dirven by diffusion and voltage driving force. *Ion pumps* in the cell membrane maintain a concentration gradient for various ions. *Ion Channels* allow charged ions to flow across the cell membrane in a highly selective way. Given a ion gradient between the inside and outside cell, a diffusion gradient is set up that causes a net flow of ions from the higher to the lower gradient. As each ion particle is charged, this causes a current flow. If a voltage potential is maintained across the membrane, positive ions will be attracted to the negative side of the cell membrane. This leads to the build up of a concentration gradient.

The Nernst potential specifies the relationship between concentration gradients and the voltage potential across the membrane:

$$E = \frac{RT}{zF} \ln \frac{[outside]}{[inside]} \qquad (7.28)$$

where $E$ is the potential, $T$ is the absolute temperature, and $R$ the gas constant, $F$ is the Faraday constant. For real cells the $E_k = [\ 70..\quad 90mv]$, $E_{Na} = [50mv]$, $E_{Ca^{2+}} = [150mv]$.

### 7.3.3 Neural Models Capture the Basic Dynamics of the Cell Body and Throw Away Some Details

The difference in neural models comes down the to the equations specifying $I$ in eqn (7.27). The trade off is between the complexity of the dynamics, the ability to analyze the systems and the ability to simulate a given model efficiently. The most well known model is the Hodgkin-Huxley model. Hodkin-Huxley empirically determined their equations (83; 82):

$$i_m = g_L(V \quad E_L) + g_k n^4(V \quad E_K) + g_{Na} m^4 h(V \quad E_{Na}) + I_s \qquad (7.29)$$

h,n,m evolve according to complex equations, $I_s$ are synaptic currents coming from other neurons. The equation can be understood in the following way: the values in parentheses $(V \quad E_x)$ are driving forces that will tend to drive the potential of the neuron to the Nerst reversal potential $E_x$. In front of each of these driving forces are weighting factors. These weighting factors determine which driving force will dominate the equation, and hence the equilibrium value for the cell membrane. If only $g_l$ is active, the cell membrane will relax toward $V \quad > E_l$. When the cell begins to fire, after the cell membrane has increased in potential, the Na part of the equation dominates and the membrane potential shoots up toward $E_{Na}$. Next the potassium term turns on and resets the neuron.

In general the integration of this equation is difficult, and the analysis is extraordinarily difficult, so researchers have turned to simplified models which capture some of the features of this system. Here we focus on computationally efficient models.

#### 7.3.3.1 Leaky Integrator

One of the simplest models is the so called leaky integrator. This model avoids the complex dynamics of firing altogether:

$$c_m \frac{dV_i}{dt} = \quad g_l V_i + I_i \qquad (7.30)$$

$$u_i = f(v_i) \qquad (7.31)$$

$$I_i = \sum_{i=1}^{N} w_{ij} u_j \qquad (7.32)$$

where the $V$ is the membrane potential, $u$ is the average firing rate of the neuron over a small time window. $f(v)$ is a function that transforms membrane voltage into firing rate.

In steady state $\frac{dV_i}{dt} = 0$. This implies that $g_l V_i = I_i$. We can often assume that $g_l = 1$. This equation is an integrator because absent the $V_i$ term, it is a perfect integrator. This equation is leaky because absent the input $I_i$, the voltage $V_i$ "leaks" back to zero.

This model does not capture the essential spiking characteristic of neurons. All precise timing information is eliminated. The next step up in complexity is an integrate and fire model.

### 7.3.3.2 Integrate and Fire

$$c_m \frac{dV_i}{dt} = g_L(V_i \quad E_L) + I_i \tag{7.33}$$

$$if(V > V_{thres}) \quad > V = V_{reset} \tag{7.34}$$

$$I_i = \sum_{i=1}^{N} w_{ij} s(u_j) \tag{7.35}$$

In the integrate and fire model, as the membrane approaches a fixed threshold, the cell 'fires' and generates a spike. It then resets the membrane voltage. This model is not sophisticated enough to build interesting models of networks controlling motors systems. We must introduce spike frequency adaptation:

$$c_m \frac{dV_i}{dt} = g_L(V_i \quad E_L) + r_m g_{sra}(V \quad E_k) + I_i \tag{7.36}$$

$$\tau_{sra} \frac{dg_{sra}}{dt} = g_{sra} \tag{7.37}$$

$$if(V > V_{thres}) \quad > V = V_{reset}, g_{sra} \quad > g_{sra} + \Delta g_{sra} \tag{7.38}$$

$$I_i = \sum_{i=1}^{N} w_{ij} s(u_j) \tag{7.39}$$

Here we have an adaptation terms. The variable $g_{sra}$ functions much like a spike averager, averaging spikes over an exponential time window. As the number of recent spike increases, the $(V \quad E_k)$ term begins to dominate. This driving force is trying to shut the neuron off. Hence with more spikes, the neurons spikes less frequently.

### 7.3.3.3 Matsuoka Oscillator

The Matsuoka Oscillator (111) another popular oscillator which seems almost like a continuous time version of the integrate and fire model with adaptation and is given as:

$$\tau_i \frac{u_i}{dt} = u_i \quad \beta f(v_i) + \sum_{j=i} w_{ij} f(u_j) + u_0 \tag{7.40}$$

$$\tau_i \frac{v_i}{dt} = v_i \quad f(u_i) \tag{7.41}$$

$$f(u) = max(0, u), (i = 1, 2) \tag{7.42}$$

As can be seen, as the neuron fires more, the value $v_i$ accumulates (almost like averaging spikes, only this neuron does not spike). As $v_i$ increases, it introduces an inhibitory term suppressing the "firing rate" of the neuron.

### 7.3.3.4 Izhikevich Model

The Hodgkin-Huxley model is capable of a rich range of behavior, but is difficult to integrate. The leaky integrator, integrate and fire with adaptation, and the Matsuoka oscillator are easy to integrate by have less richness than the HH.

The best of all-possible-world model was discovered by Izhikevich(90) :

$$\frac{dv}{dt} = 0.05v^2 + 5v + 140 \quad u + I \tag{7.43}$$

$$\frac{du}{dt} = a(bv \quad u) \tag{7.44}$$

$$if(v > 30mv)v \quad > c, u \quad > u + d \tag{7.45}$$

See Fig. 7.8 for the range of firing behaviors that can be obtained from this model. These behaviors are controlled by parameters $a, b, c$, and $d$.

## 7.3.4 Numerical Integration

Next we turn to the practical matter of computing our equations. We can do so by numerically integrating the equations on a digital computer.

The equations for neurons that we have discussed so far are of the form:

$$\frac{dx}{dt} = f(x) \tag{7.46}$$

we can write:

$$dx = f(x)dt \tag{7.47}$$

$$\Delta x \quad f(x)\Delta t \tag{7.48}$$

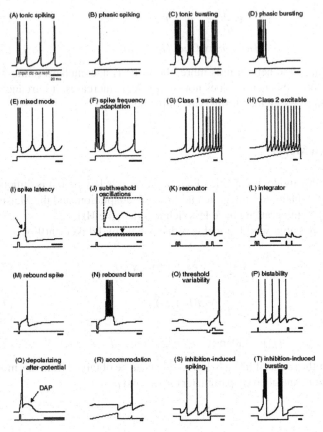

**Fig. 7.8** Range of behavior available with the Izhikevich neural model. *Electronic version of the figure and reproduction permissions are freely available at www.izhikevich.com*

$$x_{n+1} = x_n + f(x)\Delta t \tag{7.49}$$

$$x_0 = x(0) \tag{7.50}$$

The use of eqns (7.49) and (7.50) is called Eulers Integration. Derivation and use of Euler Integration is straightforward, and they are completely adequate for most equations. However equations such as the Hodgkin-Huxely equation may benefit from a better integrator such as Runge-Kutta 4th order. The iterative equations are given as:

$$x_{n+1} = x_n + \frac{h}{6}(a + 2b + 2c + d) \tag{7.51}$$

$$a = f(t_n, x_n) \tag{7.52}$$

$$b = f(t_n + \frac{h}{2}, x_n + \frac{h}{2}a) \tag{7.53}$$

$$c = f(t_n + \frac{h}{2}, x_n + \frac{h}{2}b) \tag{7.54}$$

$$d = f(t_n + h, x_n + hc) \tag{7.55}$$

Runge-Kutta is more efficient than Euler Integration. While Runge-Kutta requires more evaluations per time-step than Euler Integration, we can accurately take much larger time steps using Runge-Kutta.

However, Euler is often fast enough for small networks and has the advantage that it implementation is trivial.

### 7.3.5 Building Neural Oscillators: Nature's Coordination and Trajectory Generation Mechanism

Here we turn to the basic unit of movement generation in vertebrates: the oscillator. Oscillator circuits can be seen at the core of virtually all models of rhythmic motion generation. They are the basic building blocks of the so-called Central Pattern Generators, or CPG circuits. These circuits are groups of neurons (in vertebrates they are in the spinal cord) seen in animals capable of periodic movement including walking, running, swimming, flying etc.

**Fig. 7.9** The basic element of the central pattern generator is the Brown half-centered oscillators. Two neurons are coupled with mutual inhibition. As one neuron fires, it suppresses the other. Eventually, the firing rate wanes and the other neurons becomes active and the cycle repeats.

The basic neural unit is a network oscillator composed of two neuron in mutual inhibition, see Fig. 7.9. The basic idea of the half-centered oscillator was first proposed by Brown (37) and was the key element in a theory of central pattern generation, that is, the generation of rhythmic movement by spinal circuits without the need for a sensory trigger, or a detailed signal from the brain. While this idea was successfully repressed for period of time, in recent years it has become accepted

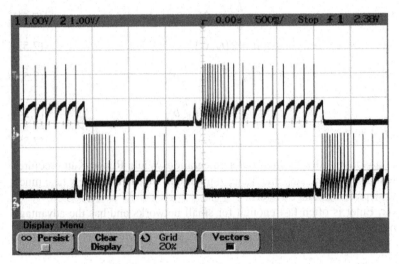

**Fig. 7.10** A Brown half centered oscillator constructed using an integrate and fire neurons with adaptation implemented with discrete components (i.e. real resistors, capacitor etc., according to the equations given above).

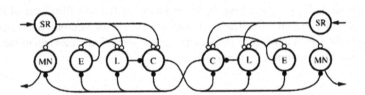

**Fig. 7.11** Basic circuitry controlling locomotion in lamprey, an eel like animal. Adapted from (61). The C neurons are coupled across the midline with mutual inhibition. They form the basis of the left-right bending of the lamprey.

since Grillner's seminal work (79; 56). In Fig. 7.11 we see an illustration of the spinal circuits for locomotion in lamprey. The lamprey is a animal that first evolved some 550 million years ago and has been relatively undifferentiated since that time. It thus gives us insight into the earliest solutions of biology. Such circuits, especially with inhibition across the mid-line to enforce a 180 degree phase shift between the left and right half body has been highly conserved. Whenever you walk, you take advantage of the lamprey solution in the left and right alternation of your legs.

## 7.3.6 Reflexes and High Level Control

The CPG circuits are modulated by sensory stimuli to adapt the CPG to the animal and to its immediate environment. Thus, the CPG is not a rigorous general given orders to a low level control system. Rather it cooperates dynamically with the environment via reflexes, for short term adaptation, and descending control from the brain, for long range anticipatory control and to achieve high level goals.

For details on the biophysics of neurons, the reader is referred to (82). For more details on neural models the reader is referred to (53)

## 7.4 Notable Systems

Here we highlight some notable models in neurobotics/biorobotics by others

1. Case Western group– Beer and colleages have created neural models of locomotion. Starting from simulation, this group has built a robot controlled by CPG based on the cockroach. This work is highly instructive. See (25; 24; 23; 22) for details on their work.
2. Neuromechanical model of swimming— Ekeberg has created a series of neuromechanical models of swimming in lamprey. What is interesting here is the integration of neurons, muscles, body and environment to achieve a complex behavior, swimming (61).
3. TAGA— Through simulation, Taga and colleagues have explored the relationship between dynamics, neural networks, and the environment. (173; 171; 170; 172)
4. Tekken— Tekken is a quadrupedal robot driven by Matsuoka oscillators. It is capable of rapid walking, and features highly innovative reflex integration with CPGs, see (94; 93).

## 7.5 GPUs

Central Processing Units have doubled in speed about every 18 months for the past 40 years (1). Graphics Processing Units have doubled in speed about every 12 months over the past 8 years (2). The result has been that graphics units are arguably the most important computational element in the modern computer.

NIVIDIA has opened their architecture to allow other to use their highly parallel architecture for general purpose simulation. GPUs are distinguished by devoting a great deal of chip real-estate to computation, instead of memory. They have a relatively small amount of high speed memory and also a large pool of They are therefore best suited for compute-bound processes. As an example, the Hodgkin-Huxley model is estimated to require 1200 flops per 1 ms of computation (91) while using only a handful of state variables. Further, neuronal models are inherently parallelizable as they rely on very similar computation applied to different state variables. On the other, hand there are some elements of the neuron elements that high dimensional and require a relatively small amount of computation for 1 ms of simulation. An example of that is the simulation of axonal delay, which can be implemented as an individual delay for each connection between an axon and a dendrites. Given the small amount of high speed memory, it is not practical to port a large and complex delay line onto GPU processors. The inhomogeneity of computational versus state space dimension for different aspects of neural simulation seems to prescribe a certain mapping of the neuronal network simulation onto a combination of GPU and CPU elements.

A neural simulation for neurons, the network connectivity, input and output provisions, and learning rules. The neuron simulation is composed of a (1) Simulation of the decision to fire, (2) Simulation of axonal delays (3) simulation of synapses (4) dendrite simulation. The learning rule is typically confined to the synapse weight. However, there is in principle no reason why axonal delays might not also be changed or altered as they might play an important role in storage of activity patterns.

The raw processing power of a GPU is staggering. The NVIDIA Telsa card with 128 core of processors is capable of nearly 500 GigaFlops of computation. A new version, to be released in late 2008 will have double that power.

Recent experiments in our lab have achieved rates of 10 million Izhikevich neurons in real-time. Currently this preliminary figure does not include long range synapses. The results are promising. Within a year, we anticipate that a computer with 4 NVIDIA GPU cards will be capable of integrating about 80 million neurons in real time, or nearly $10^8$ neurons in a single desktop computer. If GPU maintain their rate of speed increase, doubling each eary, (which is not guaranteed), within a little more than a decade it may be possible to simulate as many neurons as their are in a human brain in real-time and to control a robot.

Problems still exist for the GPU paradigm including the problem of getting information to the computational cores in an efficient manner. Yet even more insurmountable problem exist in taking advantage of the processing power we have today.

It is clear that one of the key innovations in neurorobotics will lie in harnessing GPU processing power for real-time neural computation.

## 7.6 Conclusion

Neurorobotics is a paradigm which has evolved out of the desire to understand computation in the human brain. Due to advances in the theory of neural computation, and the dramatic increase in processing power, Neurorobotics may prove to be capable of creating highly complex behaviors in life-like machines. The market acceptance of such machines is complicated and remains uncertain, but promising.

**Acknowledgements** The authors wish to thank Kiwon Sohn for his careful reading of this manuscript and useful comments as well a John Goulding for careful copies of class notes from the Spring 2008 class Neurobotics taught at the University of Arizona. The authors acknowledge the support of the University of Arizona startup fund to MAL. This chapter is derived from a lecture given by MAL at the Telluride Neuromorphic Engineering Workshop in 2008.

# Chapter 8
# Learning Inverse Dynamics by Gaussian Process Regression under the Multi-Task Learning Framework

Dit-Yan Yeung and Yu Zhang

**Abstract** In this chapter, dedicated to Dit-Yan's mentor and friend George Bekey on the occasion of his 80th birthday, we investigate for the first time the feasibility of applying the *multi-task learning* (or called *transfer learning*) approach to the learning of inverse dynamics. Due to the difficulties of modeling the dynamics completely and accurately and solving the dynamics equations analytically to obtain the control variables, the machine learning approach has been regarded as a viable alternative to the robotic control problem. In particular, we learn the inverse model from measured data as a regression problem and solve it using a nonparametric Bayesian kernel approach called *Gaussian process regression* (GPR). Instead of solving the regression tasks for different degrees of freedom (DOFs) separately and independently, the central thesis of this work is that modeling the inter-task dependencies explicitly and allowing adaptive transfer of knowledge between different tasks can make the learning problem much easier. Specifically, based on data from a 7-DOF robot arm, we demonstrate that the learning accuracy can often be significantly increased when the multi-task learning approach is adopted.

## 8.1 Appreciation and Dedication

When Dit-Yan arrived at the University of Southern California (USC) in the mid 1980s, he was planning to do theoretical research on the models of computation, possibly including computational and mathematical models for human intelligence. Robotics was initially not in his mind. Shortly afterwards he learned of the interest-

Dit-Yan Yeung
Department of Computer Science and Engineering, Hong Kong University of Science and Technology, Clear Water Bay, Hong Kong, China, e-mail: dyyeung@cse.ust.hk

Yu Zhang
Department of Computer Science and Engineering, Hong Kong University of Science and Technology, Clear Water Bay, Hong Kong, China, e-mail: zhangyu@cse.ust.hk

131

ing robotics research carried out in George Bekey's laboratory and was fascinated by the fun and challenges in this research area. It was fortunate that George Bekey agreed to be his PhD advisor.

George has always been a very open-minded scholar who gives unfailing support for new ideas and explorations, no matter how "silly" they first appear. Dit-Yan was given a lot of freedom and encouragement to explore new ideas and even unexplored territories. Eventually he worked on machine learning and robotic control for his doctoral research. Although for various reasons he no longer worked on robotics after graduation, the research experience he gained from his doctoral study helped him greatly in pursing his current research interests in machine learning and pattern recognition. In psychology such learning experience is known as *transfer of learning*. This notion has inspired the development of an active research topic in machine learning, known as *transfer learning* or *multi-task learning*. In this chapter, it carries a very special meaning to put the three themes, namely, *machine learning*, *robotic control*, and *multi-task learning*, together as an appreciation and dedication to George for his great and fatherly mentorship.

The second author of this chapter is one of the recent PhD students of Dit-Yan. So this *transfer* originated from George will continue on and on. Thank you George!

## 8.2 Robotic Control

### 8.2.1 Kinematics and Dynamics

Kinematics and dynamics are two important aspects that are central to the control of robots or articulated objects with jointed rigid segments (125; 26). For given angles of the joints, the *forward kinematics* problem refers to the computation of the position and orientation of the end effector of a robot. The more difficult problem is the inverse problem, called *inverse kinematics*, which determines the joint angles of the robot in order for its end effector to achieve some desired pose.

To control the movement of a robot, we have to consider its dynamics as well in addition to the kinematics. The *forward dynamics* problem refers to the computation of the trajectory in terms of the joint angles, velocities and accelerations given the torques at the joints. *Inverse dynamics*, like inverse kinematics, is the inverse problem which is much more difficult to solve than the forward problem.

### 8.2.2 Reasons Against Analytic Solutions

Analytic solutions for the kinematics and dynamics equations are often quite expensive to obtain. The main difficulties arise from the strong coupling between different degrees of freedom (DOFs) and the high dimensionality and nonlinearity of the equations for robots with many DOFs, such as humanoid robots which have aroused a great deal of interests in the robotics community over the past decade. A robot is a complex system whose parameters are not constant but can vary with the control or state variables. Also, some parameters may never be known precisely for control purposes. Methods such as adaptive control using parameter identification techniques exist for solving such problems. However, these methods usually require a complete model of the system in order to identify the parameters. In practice, obtaining an accurate model is very difficult, if not totally impossible.

## 8.2.3 Insights from Human Arm Control

In human arm control, it is very unlikely for some dynamics equation to be solved analytically somewhere inside the brain to issue control commands to the arm. To move from one location to the other, the arm is usually controlled to go through an initial *feedforward* phase of fast motion which brings the arm to the right "ballpark" of the desired location (14). This is then followed by a second phase of fine motion control which relies heavily on sensory *feedback*. The initial phase does not require high precision in position control. Rather, its primary concern is to provide fast computation of control commands for moving the arm rapidly to some neighborhood of the desired location. Such a two-phase scheme is also useful for robotic arm control. We only need an approximate model for control in the feedforward path. Internal and external sensors can then be used to provide feedback information to correct the errors made by the feedforward model.

## 8.2.4 Learning and Control

The considerations above motivate robotics researchers to take a machine learning approach as a viable alternative to the robotic control problem. As part of his doctoral thesis research, the first author of this chapter proposed a neural network model called *context-sensitive network* (200; 199) as a machine learning approach to robotic control. Since then, the learning approach has become more commonly used especially when more complex robotic systems such as humanoid robots are studied. The focus of this chapter is on a learning approach to the inverse dynamics problem.

## 8.3 Learning Inverse Dynamics

The most common learning approach to the inverse dynamics problem is to learn the inverse model from measured data as a *regression* problem. Since typical robotic systems have multiple DOFs, the regression problem involves multiple response variables. For example, each response variable corresponds to the torque at one joint.

### 8.3.1 Recent Work

Nonparametric regression methods are more suitable for solving the inverse dynamics problem due to their higher model flexibility. The method known as *locally weighted projection regression* (LWPR) (187; 186; 185) is currently the standard learning method used in the robotics community since it is capable of online, real-time learning even for complex robots such as humanoid robots. However, many other powerful regression methods have been developed in the machine learning community over the past decade or so. In particular, the *kernel approach* (153) is arguably the most popular due to its mathematical elegance as well as promising performance in practice. *Support vector regression* (SVR) is an extension of the support vector machine (SVM) from classification problems to regression problems (183; 153). Besides, *Gaussian process* (GP) models (140) are nonparametric Bayesian kernel machines that, like SVR, have demonstrated state-of-the-art performance in many regression applications. Recently, an empirical performance comparison was conducted to compare LWPR, SVR and GP regression (GPR) for learning inverse dynamics (119; 120). While LWPR is generally more efficient, SVR and GPR are more accurate and have fewer hyperparameters to set. LWPR has many meta parameters which are tedious to tune.

### 8.3.2 Learning Inverse Dynamics as a Regression Problem

The general form of the dynamics equation can be expressed as

$$\tau = \mathbf{M}(\theta)\ddot{\theta} + \tau_v(\theta,\dot{\theta}) + \tau_g(\theta) + \tau_f(\theta,\dot{\theta}), \tag{8.1}$$

where $\theta$, $\dot{\theta}$ and $\ddot{\theta}$ are the joint angles, velocities and accelerations, respectively, $\tau$ is the torque vector, $\mathbf{M}$ is the mass or inertia matrix, $\tau_v$ is a vector of centrifugal and Coriolis terms, $\tau_g$ is a vector of gravity terms, and $\tau_f$ is a vector of friction terms. By regarding the learning of the inverse dynamics as a regression problem, we rewrite (8.1) as the following regression function

$$\tau = \mathbf{g}(\theta,\dot{\theta},\ddot{\theta}), \tag{8.2}$$

where $\theta$, $\dot{\theta}$ and $\ddot{\theta}$ represent the explanatory variables (independent variables) and $\tau$ represents the response variables (dependent variables).[1] Note that $\mathbf{g}$ is a vector function since $\tau$ is multivariate. For example, if the robot is a 7-DOF arm, then there are 21 independent variables and 7 dependent variables.

## 8.4 Gaussian Process Regression

In this section, we first review the GP approach to regression. We then report our experimental results on using GPR for learning inverse dynamics.

### 8.4.1 Brief Review

We first describe a *weight-space view* on GPR. Consider a regression problem with $p$ input variables represented as a $p$-dimensional input vector $\mathbf{x} \quad \mathbb{R}^p$ and an output variable represented as a scalar output value $y \quad R$. We are given a training set $\mathscr{D} = (\mathbf{x}_i, y_i) \; {}_{i=1}^n$ of $n$ observations in the form of input-output pairs. Let $\phi(\ )$ denote a $d$-dimensional vector function representing $d$ fixed basis functions that transform $\mathbf{x}$ (usually nonlinearly) from the input space to some other space. The standard linear regression model with Gaussian noise is given by

$$f(\mathbf{x}) = \mathbf{w}^T \phi(\mathbf{x}) \tag{8.3}$$

$$y = f(\mathbf{x}) + \varepsilon, \tag{8.4}$$

where $\mathbf{w} \quad \mathbb{R}^d$ is a weight vector, $f(\ )$ is a latent function, and $\varepsilon$ is an independent and identically distributed (i.i.d.) Gaussian noise variable with zero mean and variance $\sigma^2$, i.e., $\varepsilon \quad N(0, \sigma^2)$.

Uncertainty is modeled probabilistically by defining $\mathbf{w}$ to be a random variable following a multivariate Gaussian distribution with zero mean and covariance matrix $\Sigma$, i.e., $\mathbf{w} \quad N(\mathbf{0}, \Sigma)$. The prior distribution over $\mathbf{w}$ induces a corresponding prior distribution over $f(\mathbf{x})$. Let us define $\mathbf{f} = (f(\mathbf{x}_1), \dots, f(\mathbf{x}_n))^T$ and $\Phi = (\phi(\mathbf{x}_1), \dots, \phi(\mathbf{x}_n))^T$. Thus the training data set $\mathscr{D}$ can be expressed by the linear regression model in matrix form as

$$\mathbf{f} = \Phi \mathbf{w}. \tag{8.5}$$

We are interested in the joint distribution of the function values.

A GP is a collection of random variables such that any finite number of them exhibit a consistent joint Gaussian distribution. GPR is a nonparametric Bayesian

---

[1] Strictly speaking the variables $\ddot{\theta}$ in the regression problem are not exactly the joint accelerations, but they are corrected by the closed-loop results to give the joint accelerations. See (120) or (46) for more discussions on inverse dynamics control.

approach to regression which assumes that the latent function $f$ follows a GP prior (a prior distribution over functions, which in general are infinite-dimensional objects). Since $\mathbf{f}$ is a linear combination of Gaussian random variables, $\mathbf{f}$ is itself Gaussian with the following mean vector and covariance matrix:

$$\mathbb{E}[\mathbf{f}] = \Phi\mathbb{E}[\mathbf{w}] = \mathbf{0} \tag{8.6}$$

$$\text{cov}[\mathbf{f}] = \mathbb{E}[\mathbf{ff}^T] = \Phi\mathbb{E}[\mathbf{ww}^T]\Phi^T = \Phi\Sigma\Phi^T. \tag{8.7}$$

So we have

$$\mathbf{f} \sim N(\mathbf{0}, \Phi\Sigma\Phi^T). \tag{8.8}$$

Instead of proceeding with the weight-space view, the *function-space view* bypasses the modeling of the weights $\mathbf{w}$ and the basis functions $\phi(\ )$. Based on this view, the Gaussian distribution for $\mathbf{f}$ is directly modeled as

$$\mathbf{f} \sim N(\mathbf{0}, \mathbf{K}), \tag{8.9}$$

where $\mathbf{K}$ is a positive semidefinite matrix with elements $\mathbf{K}_{ij} = k(\mathbf{x}_i, \mathbf{x}_j)$ for some covariance function $k(\ ,\ )$, which corresponds to a Mercer kernel in the kernel approach (153). Like the kernel approach in general, one advantage of this function-space view is that we may choose a covariance function $k(\ ,\ )$ that corresponds to using a very large or even infinite number of basis functions giving high expressiveness. Since $\varepsilon$ is an additive i.i.d. noise term, we can easily show that

$$\mathbf{y} \sim N(\mathbf{0}, \mathbf{K} + \sigma^2\mathbf{I}), \tag{8.10}$$

where $\mathbf{y} = (y_1, \ldots, y_n)^T$ and $\mathbf{I}$ is the $n \times n$ identity matrix.

For a new test case $\mathbf{x}_*$, the predictive distribution of its latent function value $f(\mathbf{x}_*)$ has the following Gaussian distribution:

$$p(f(\mathbf{x}_*) \mid \mathbf{x}_*, \mathscr{D}) = N(\mathbf{k}_*^T\mathbf{C}^{-1}\mathbf{y}, k(\mathbf{x}_*, \mathbf{x}_*) - \mathbf{k}_*^T\mathbf{C}^{-1}\mathbf{k}_*), \tag{8.11}$$

where $\mathbf{k}_* = (k(\mathbf{x}_1, \mathbf{x}_*), \ldots, k(\mathbf{x}_n, \mathbf{x}_*))^T$ and $\mathbf{C} = \mathbf{K} + \sigma^2\mathbf{I}$. Computing this distribution requires inverting the $n \times n$ matrix $\mathbf{C}$ with $O(n^3)$ time complexity.

### 8.4.2 Gaussian Process Regression for Learning Inverse Dynamics

The GPR method reviewed above assumes that the regression function is univariate. However, for learning inverse dynamics, usually there are multiple DOFs and hence $\mathbf{g} = (g_1, \ldots, g_m)$ in (8.2) is a vector function. Learning $\mathbf{g}$ may be achieved by learning each of the component functions $g_j$ separately and independently, as in (119; 120). In this subsection, we report some experimental results we have obtained based on this setting. This naïve setting will be extended in the next section by regarding the learning of different DOFs as dependent tasks via sharing information among them.

We use GPR to learn the inverse dynamics of a 7-DOF SARCOS anthropomorphic robot arm by using the data set in http://www.gaussianprocess.org/gpml/data/. Each observation in the data set consists of 21 input features (7 joint positions, 7 joint velocities, and 7 joint accelerations) and the corresponding 7 joint torques for the 7 DOFs. There are two disjoint sets, one for training and one for testing. We only use the first 10000 examples of the training set for training but the entire test set for testing. We use the MATLAB code provided by Rasmussen and Williams in http://www.gaussianprocess.org/gpml/code/gpml-matlab.zip for performing GPR. For our performance measure, like in (119), we adopt the normalized mean squared error (nMSE) which is defined as the mean squared error divided by the variance of the target. The squared exponential covariance function is used for the GP.[2] We want to see how the training sample size affects the learning accuracy. To do so, we perform experiments by gradually increasing the training sample size from 100 to 1100 by 100 at a time. For each sample size, multiple runs are performed on different training sets of the same size and the average nMSE is reported. We also perform an experiment once on all the 10000 training examples.

The results are depicted in Figures 8.1–8.7. Each figure shows the regression result of one DOF expressed in terms of the average nMSE under varying sample size. Ideally one would expect the average nMSE to decrease when the sample size is increased. However, this trend is not clearly observed and the variation of nMSE is only within a very small range. Essentially we can conclude that increasing the sample size beyond 100 does not significantly increase the learning accuracy. Nevertheless, the primary objective of this paper is on comparing the standard GPR with a multi-task extension which will be studied in the next section.

## 8.5 Multi-Task Gaussian Process Regression

A common daily experience is that learning to solve a problem can be much easier if we have learned to solve a different but related problem before. The more related the two tasks are, the more we can benefit from the previous learning experience. This is related to the notion of *transfer of learning* (62) in psychology. In machine learning this is known as *multi-task learning, transfer learning, inductive transfer,* or *learning how to learn* (175; 19; 20), which has received a lot of attention in the machine learning community over the past decade or so.

In order that we can benefit from the multi-task learning setting, some tasks to learn must be related in some sense and some common information must be shared among these related tasks. One natural approach to this learning problem is through hierarchical Bayesian modeling. While classical Bayesian modeling is based on parametric models, nonparametric Bayesian models are generally more desirable due to their higher model flexibility. GP is a promising nonparametric Bayesian ap-

---

[2] The squared exponential covariance function is also known as the radial basis function (RBF) or Gaussian covariance function.

proach and hence our focus here will be on multi-task learning based on the GP approach (115; 102; 154; 202; 32; 33; 201).

Note that transferring the learning experience regardless of whether the tasks are related or not may lead to impaired performance. Ideally, how much to transfer should depend on the "relatedness" between tasks. Many multi-task learning methods studied in the past simply assume *a priori* that the tasks concerned are related. This assumption may be too strong for some real-world applications. A promising method was proposed recently for multi-task learning by modeling the inter-task dependencies explicitly (33). In so doing, transfer can be made *adaptive* in the sense that the degree of transfer can depend on how related the tasks are. We will briefly review this method below and then apply it to the learning of inverse dynamics. The objective is to demonstrate that the learning problem can be made much easier under the multi-task learning framework by sharing the learning experience among different tasks.

### 8.5.1 Brief Review of Bonilla et al.'s Method (33)

The training set is now represented as $\mathscr{D} = \{(\mathbf{x}_i, \mathbf{y}_i)\}_{i=1}^n$ where $\mathbf{x}_i \in \mathbb{R}^p$ and $\mathbf{y}_i = (y_{i1}, \ldots, y_{im})^T \in \mathbb{R}^m$. For the $k$th task, there is a corresponding latent function $f_k$. We assume that $f_k(\mathbf{x}_i)$ has a GP prior with zero mean and each entry of the covariance matrix is

$$\mathbb{E}[f_k(\mathbf{x}_i) f_l(\mathbf{x}_j)] = \mathbf{K}_{kl}^f k^x(\mathbf{x}_i, \mathbf{x}_j), \qquad (8.12)$$

where $\mathbf{K}^f$ is an $m \times m$ symmetric, positive semidefinite matrix with each entry $\mathbf{K}_{kl}^f$ specifying the similarity between tasks $k$ and $l$, and $\mathbf{K}^x$ is an $n \times n$ symmetric, positive semidefinite matrix with each entry defined by $k^x(\cdot, \cdot)$, which is a covariance function over inputs just like $k(\cdot, \cdot)$ in Section 8.4. If we define $\mathbf{f} = (f_{11}, \ldots, f_{n1}, f_{12}, \ldots, f_{n2}, \ldots, f_{1m}, \ldots, f_{nm})^T$, we can immediately see that

$$\mathbf{f} \sim N(\mathbf{0}, \mathbf{K}^f \otimes \mathbf{K}^x), \qquad (8.13)$$

where $\otimes$ denotes the Kronecker product. We also assume that each task $k$ has a separate additive noise term, i.e.

$$y_{ik} = f_k(\mathbf{x}_i) + \varepsilon_k, \qquad (8.14)$$

where $\varepsilon_k \sim N(0, \sigma_k^2)$ or, equivalently, $y_{ik} \sim N(f_k(\mathbf{x}_i), \sigma_k^2)$.

For a new test case $\mathbf{x}_*$ that belongs to task $k$, which is one of the $m$ tasks in the training data, the predictive distribution of $f_k(\mathbf{x}_*)$ can be expressed in a form similar to that in (8.11) for the univariate (single-task) case, with its mean prediction given by

$$\bar{f}_k(\mathbf{x}_*) = (\mathbf{k}_k^f \otimes \mathbf{k}_*^x)^T \mathbf{C}^{-1} \mathbf{y} \qquad \mathbf{C} = \mathbf{K}^f \otimes \mathbf{K}^x + \mathbf{D} \otimes \mathbf{I}, \qquad (8.15)$$

where $\mathbf{k}_k^f$ denotes the $k$th column of $\mathbf{K}^f$, $\mathbf{k}^x = (k^x(\mathbf{x}_1, \mathbf{x}\ ), \ldots, k^x(\mathbf{x}_n, \mathbf{x}\ ))^T$ is the vector of covariances between the $n$ training data points and the test case $\mathbf{x}\ $, $\mathbf{y} = (y_{11}, \ldots, y_{n1}, y_{12}, \ldots, y_{n2}, \ldots, y_{1m}, \ldots, y_{nm})^T$, and $\mathbf{D}$ is an $m\ \ m$ diagonal matrix with $\mathbf{D}_{kk} = \sigma_k^2$. The covariance matrix can be similarly generalized from that in (8.11). Note that $\mathbf{C}$ is of size $mn\ \ mn$ which is much larger than before.

While $\mathbf{K}^x$ is modeled *parametrically* via some kernel parameters $\theta_x$ of the kernel function $k^x(\ ,\ )$, $\mathbf{K}^f$ is modeled in a *nonparametric* manner. Hence $\theta_x$ and $\mathbf{K}^f$ need to be determined, either manually or, preferably, automatically. To distinguish them from the (latent) weight parameters $\mathbf{w}$ of the model itself, these parameters are often referred to as hyperparameters in Bayesian modeling. A method was proposed in (33) for learning these hyperparameters from data. In addition, they also proposed a method for dealing with the problem of large $n$. We refer the readers to their paper for more details.

### 8.5.2 Multi-Task Gaussian Process Regression for Learning Inverse Dynamics

As in Section 8.4.2, the learning of each function $g_j$ corresponds to one learning task. However, the difference here is that we now learn the inter-task dependencies explicitly and make use of them in learning the tasks to achieve multi-task learning.

The experimental settings are the same as those in Section 8.4.2. The results are depicted in Figures 8.8–8.14. For the convenience of comparing the performance of GPR and Multi-Task GPR, we also incorporate the results from Figures 8.1–8.7 to Figures 8.8–8.14. From the results, we can see that the performance of Multi-Task GPR is usually significantly better than that of GPR, except for the 6th DOF. In fact, the performance of Multi-Task GPR with as few as 100 training examples is often much better than that of GPR with as many as 10000 training examples, demonstrating the effectiveness of multi-task learning. A possible explanation for the somewhat abnormal behavior of the 6th DOF is the improper characterization of inter-task similarity which makes the improper transfer to impair the learning performance. Further investigation is needed to fully unveil the truth.

## 8.6 Conclusion

In this chapter, we have presented our first attempt to investigate the feasibility of applying the multi-task learning approach to robotic control. Although our experimental investigation is preliminary due to the limit of time, the results obtained are very encouraging. Specifically, we demonstrate that by adopting the multi-task learning approach, the learning accuracy can be significantly improved even using two orders of magnitude fewer training examples than that reported recently by others.

There exist many interesting opportunities to take this work forward. Besides the computational and algorithmic issues related to the multi-task learning method itself, more extensive experimental investigation on the robotic control problem is necessary. Among other things, combining feedforward nonlinear control with inverse dynamics control for the real-time control of humanoid robots is a challenging yet rewarding research problem to pursue.

**Acknowledgements** This research has been supported by General Research Fund 621407 from the Research Grants Council of the Hong Kong Special Administrative Region, China.

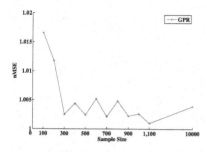

**Fig. 8.1** nMSE of GPR on the 1st DOF under varying sample size.

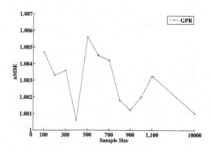

**Fig. 8.2** nMSE of GPR on the 2nd DOF under varying sample size.

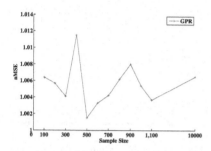

**Fig. 8.3** nMSE of GPR on the 3rd DOF under varying sample size.

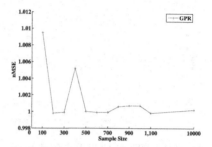

**Fig. 8.4** nMSE of GPR on the 4th DOF under varying sample size.

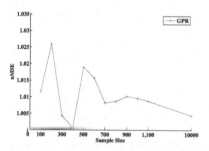

**Fig. 8.5** nMSE of GPR on the 5th DOF under varying sample size.

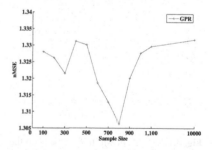

**Fig. 8.6** nMSE of GPR on the 6th DOF under varying sample size.

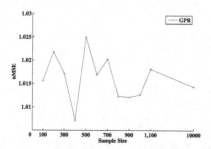

**Fig. 8.7** nMSE of GPR on the 7th DOF under varying sample size.

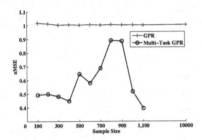

**Fig. 8.8** nMSE of Multi-Task GPR on the 1st DOF under varying sample size.

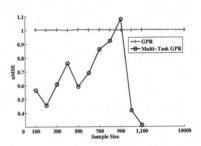

**Fig. 8.9** nMSE of Multi-Task GPR on the 2nd DOF under varying sample size.

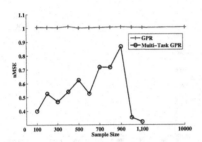

**Fig. 8.10** nMSE of Multi-Task GPR on the 3rd DOF under varying sample size.

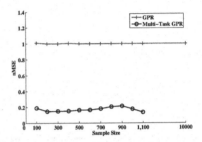

**Fig. 8.11** nMSE of Multi-Task GPR on the 4th DOF under varying sample size.

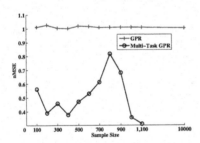

**Fig. 8.12** nMSE of Multi-Task GPR on the 5th DOF under varying sample size.

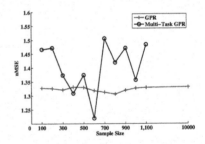

**Fig. 8.13** nMSE of Multi-Task GPR on the 6th DOF under varying sample size.

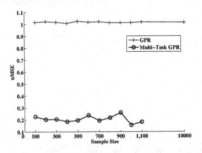

**Fig. 8.14** nMSE of Multi-Task GPR on the 7th DOF under varying sample size.

# Part II
# Tributes and Recollections from Former Students

In the course of his long and distinguished career, Professor Bekey had many associates. It is perhaps not unfair to say that some of the strongest bonds he formed were with his graduate students, many of whom stay in regular touch with him (and with each other) to this day. This part of the book gives a brief survey of Professor Bekey's career and an account of his students (the Bekey 'Tribe'). It also includes some tributes and recollections by his former students.

# Chapter 9
# Professor George Albert Bekey

Gaurav S. Sukhatme

To synopsize Professor Bekey's long and illustrious career is tricky. There are many stories to tell, many kudos to list, and, for the reader, many lessons to learn. A comprehensive account will have to wait for a biographer; what follow are selected highlights, illustrative of Professor Bekey's nature and professional achievements.

## 9.1 Personal Life

Professor George Bekey was born in Bratislava in 1928. His first given name was Juraj. This changed to Jiri when the family moved to Prague in 1931, and then to Jorge when they moved to Bolivia in 1939 as a consequence of the worsening political climate across Europe. Their exit from Czechoslovakia was dramatic and timely. Professor Bekey's grandparents did not leave the country and he was never to see them again. In 1945, the family moved to Los Angeles, where, at the age of 17, he became George. Professor Bekey's father was an Electrical Engineer, and his mother was a chemistry graduate, an unusual achievement for a woman in the first quarter of the 20th century. He has an older brother, Ivan Bekey, an expert in space systems, who spent his professional life at NASA.

Professor Bekey is married to Dr. Shirley Bekey, a child psychologist. They have two children, Dr. Ron Bekey, a Professor at Portland Community College, and Michelle Bekey, a principal at Bekey and Company which provides strategic communications consulting.

---

Gaurav S. Sukhatme
Department of Computer Science, University of Southern California e-mail: gaurav@usc.edu

## Analyzer Interconnections for Direct Determination of Power-System Swing Curves

G. A. BEKEY      F. W. SCHOTT
ASSOCIATE MEMBER AIEE   ASSOCIATE MEMBER AIEE

**T**HE calculation of power-system swing curves is generally performed by using one of several point-by-point techniques in conjunction with an a-c network analyzer. The technique is well known and described in the standard reference texts on the subject,[1,2,3] and while greatly simplified by the network analyzer, is still laborious and time-consuming. At least one analog has been devised[4] for the automatic and continuous determination of swing curves, and several have been proposed.[4-7]

This paper presents a development of the method for direct determination of swing curves by the interconnection of a network analyzer with a differential analyzer, substantially as proposed by Kimbark.[5] The network analyzer simulates the electrical network, while the differential-analyzer computer is arranged to represent the mechanical inertias of the generators. In this paper, which attempts to illustrate the method rather than to exploit it, the technique is applied to the simple case of a single generator connected to an infinite bus, allowing comparative accuracies with analytical and point-by-point methods to be discussed. Suggestions for possible extension of the technique to other problems and utilization of more readily available computers are discussed in the concluding sections of the paper.

### Determination of Swing Curves

The oscillation of a generator rotor in transient stability analysis is ordinarily determined by integration of the swing equation

$$\frac{d^2\delta_i}{dt^2} = \frac{180f}{H_i}(T_{mi}-T_{ei}) \qquad (1)$$

where

$\delta_i$ = power angle, degrees
$f$ = frequency, cycles per second
$H_i$ = inertia constant, kw-seconds per kva
$T_{mi}$ = mechanical driving torque, per unit
$T_{ei}$ = electrical output torque, per unit
$t$ = time, seconds

The subscript $i$ is used to emphasize

the fact that equation 1 must be applied simultaneously to each machine constituting the system. In simple systems, the equal-area criterion may be utilized[3] or generalized solutions may be applied.[3] In more complex cases, the solution to equation 1, which may be written directly as

$$\delta_i(t) = \frac{180f}{H_i}\int_0^t \int_0^t (T_{mi}-T_{ei})dt\,dt \qquad (2)$$

is built up by approximating the integral by a finite sum. In this step-by-step method, the network analyzer can be used to provide the successive values of $T_{ei}$, since $T_{ei}$ is substantially equal to the generator power output, while $T_{mi}$ is generally assumed to be constant, or at most a function of the independent variable, time. At the end of every step, each rotor angle $\delta_i$ is adjusted by the calculated increment.

In the method of this paper, the solution to equation 1 is again carried out by direct integration, as indicated by equation 2. Here, however, the integration is carried out continuously, using the differential analyzer, thus eliminating the errors of the step-by-step method, while also reducing the time required for roughly a factor of ten. Furthermore, each swing curve is presented directly on a differential-analyzer output table during the computation.

Fig. 1 illustrates, in block diagram form, the essential features of the analyzers and their interconnection. The electrical power supplied to the network is metered at the generator wattmeter and converted by an operator to shaft rotation. Synchros are used to transmit this rotation to the differential analyzer, which solves equation 2 with the use of two integrator units and appropriate gear trains. The power angle obtained from this computation is then fed back continuously to the generator rotor of the network analyzer. Plots of actual swing curves (angle-time curves) are obtained on differential-analyzer output tables.

For each generator in the electrical system an interconnection, as of Fig. 1,

is needed. Hence, a differential analyzer with 14 integrator units permits the solution of a 7-machine system.

### Illustrative Examples

The system of Fig. 1 was employed for the solution of two simple problems in order to test its accuracy by comparison with known solutions. The first problem consisted of a single generator connected to an infinite bus through a reactance, as illustrated in Fig. 2. If the mechanical input to the generator is suddenly reduced to zero this system will oscillate. The variation in rotor angle may be computed analytically as shown in the Appendix. The calculated swing curve is compared with that obtained by the analyzers on Fig. 3, showing agreement within 2 per cent.

The second case shows the effect of interval size in the point-by-point method of solution. As would be expected, since the present computer solution is continuous rather than stepwise, the cumulative errors of the stepwise technique should be minimized. The system is illustrated in Fig. 4, showing the point where the fault is applied. The fault is cleared by simultaneous operation of the breakers on the lower transmission line in 0.2 second. The analyzer solutions are compared with the hand-computed solutions in Fig. 5.

The three curves illustrated compare the continuous solution to that obtained stepwise using intervals of 0.05 and 0.01 second. It readily can be seen that in this step-by-step problem, the cumulative errors of the stepwise solution actually suggest an unstable system, while a smaller interval or the continuous solution point to a stable system, as verified by the equal-area criterion. Agreement of successive solutions is within 2 per cent.

### Limitations and Advantages

It is quite clear that the success of this interconnection is limited to network analyzers with low-regulation generators.

Paper 54-14, recommended by the AIEE Transmission and Distribution Committee and approved by the AIEE Committee on Technical Operations for presentation at the AIEE Winter General Meeting, New York, N. Y., January 18-22, 1954. Manuscript submitted October 19, 1953; made available for printing November 12, 1953.

G. A. BEKEY and F. W. SCHOTT are with the University of California, Los Angeles, Calif.

The authors wish to express their appreciation to Don Lebell of the Department of Engineering and Thomas O. Ellis of the Rand Corporation for their advice and assistance in the early stages of the work. Thanks are also due to M. R. Davis and E. C. DeLand of the differential analyzer staff at the University of California, Los Angeles, for their able assistance in the experimental work.

**Fig. 9.1** Professor Bekey's first paper (April 1954).

# 9.2 Research

To many who don't work at a University or know anyone who does, what Professors do is largely a mystery. Professors teach, but what else? At a major research university faculty spend a substantial portion of time engaged in fundamental research and advising PhD students. Professor Bekey spent his professional life as a member of the faculty of the University of Southern California (USC). Over a period of 40 years, his research spanned eight broad areas, each a field of inquiry in its own right; his contributions to these areas are summarized below.

**Application of Computers to Unusual Domains:** While still a graduate student at UCLA, Professor Bekey succeeded in connecting a mechanical analog computer to a static electrical power system simulator, in order to solve the differential equations of power system oscillation. Since these computers were located on different floors of a building, this required (among other things) running cables through a hole in the floor. This work resulted in his first journal publication, in April 1954 (see Figure 9.1). Also as part of his early work, he was the first to solve the Hodgkin-Huxley equations of nerve propagation on an analog computer.

**Mathematical Models of Human Operators:** Professor Bekey's Ph.D. thesis was concerned with development of a discrete-time model of a human operator in a closed-loop control system. Over a period of some twenty years he and his students developed mathematical models to characterize the behavior of human operators in control systems under a variety of conditions. His work in this area focused on human behavior with tactile inputs, and the mathematical representation of the behavior of human drivers in single lane car following. It is interesting that this work preceded (and in many ways anticipated) the current interest in automated vehicles and highways by many years.

**Optimization by Random Search:** Professor Bekey developed several algorithms for optimization of dynamic systems by random search. This was first published in 1966, and later expanded and improved in 1980 in collaboration with Professor Sami Masri of the USC Civil Engineering department. Jointly they applied these methods to a variety of structural problems, such as earthquake-induced building oscillations, which resulted in a number of papers.

**Biology and Medicine:** Throughout his academic career, Professor Bekey has had an interest in applying engineering and computer science methods to understanding biological phenomena. This work has included modeling respiration in the fetus and adult, understanding muscle function, modeling human gait and using expert systems to diagnose its malfunctions. Many of these papers were published in physiological journals.

**System Identification and other Control System Contributions:** During the first twenty years of his academic career, the estimation of system parameters from input-output measurements was Professor Bekey's major research interest. Much of his work on human operator models provided the data for new identification algorithms.

**Multifingered Robot Hands:** Professor Bekey (and his long-time collaborator from the former Yugoslavia, Professor Rajko Tomovic) designed and built a number of multifingered hands for robot manipulators, and developed a theory for the representation of robot grasping. They also developed an approach to minimal complexity control of prosthetic hands. In this work Professor Bekey also had a long-standing collaboration with Professor Michael Arbib, who he recruited to USC in the mid 1980s.

**Manufacturing:** Professor Bekey's work on manufacturing concentrated on issues of assembly and design-for-assembly, including the use of artificial intelligence methods for re-design and interleaving of planning and design. He recruited Professor Aristides Requicha to USC in this area in the mid 1980s, and also collaborated with Professor Behrokh Knoshnevis extensively in this area. Another recruit he brought to USC was Professor Ken Goldberg (now at UC Berkeley).

**Autonomous Mobile Robots:** During the 1980s and 1990s Professor Bekey and his students concentrated on the theory and development of autonomous mobile robots. Among these vehicles were a walking machine which used genetic algorithms to evolve suitable gait patterns, and an autonomous helicopter. This phase of Professor Bekey's professional life lasted over two decades and motivates most of the work described in the first part of this book.

## 9.3 Teaching and Students

Professor Bekey graduated 38 PhDs during his career, nearly one each year on the average. 17 of these went on to careers in academia, the others have risen to leadership positions in research labs and industry around the world. The "Bekey Tribe" is comprised of a collection of Professor George Bekey's former students, three students of the late legendary Professor Richard Bellman, and associates. In the following chapter, two senior members of the tribe tell its story.

Being George's PhD student was an interesting experience. He wanted students to develop their own interests, and while he counselled, he never forced students to do what he wanted. He showed unflagging support and belief in his students when they needed it most. You forgot when you worked with him that he was an academic giant. Some of the personal tributes and recollections that follow in a later chapter speak to these qualities in a little more detail.

## 9.4 Service to the University and the Profession

While he was working on this research, Professor Bekey was steadily climbing the academic ladder. He was granted tenure in 1964 (only two years after joining USC), and promoted to full Professor in 1968. He was instrumental in founding the USC Biomedical Engineering department. From 1978 to 1982 he was chair of the USC Electrical Engineering-Systems department, and from 1984-89 he was chair of the USC Computer Science department. He founded the USC Robotics Research Lab in 1983 and led it for 15 years. From 1995 to 1999 he served as the Associate Dean for Research in the USC School of Engineering (now the USC Viterbi School of Engineering).

The following story is illustrative of Professor Bekey's foresight and ability to make connections. In 1972, Keith Uncapher, then at RAND Corporation, impressed officials at the Advanced Research Projects Agency (ARPA) by his work on packet switching for computer communication networks. ARPA wanted Uncapher to quickly grow his research. However the senior management at RAND responded coolly to the proposal. Uncapher had an assurance from ARPA that his research would get support wherever he located it, so he approached UCLA, and was told he'd have to wait at least 15 months. 'I don't have three weeks,' Uncapher is said to have replied at the time.

This much is well known. What is less well known is that Professor Bekey put Keith Uncapher in touch with the then-dean of the School of Engineering, Zohrab Kaprielian, a legend for his ability to seize opportunities. Dean Kaprelian and Uncapher met on a Monday and closed the deal by Friday, deciding to form the the the USC Information Sciences Institute (ISI), and the rest is history. Since its inception, ISI has been credited with enabling the Internet to grow from a research system to a major national and international infrastructure.

What is extraordinary about Professor Bekey's career is that such examples are easy to find. This particular one had an immense impact, others were less influential, but all had the same driving motivation behind them. When he saw an excellent new idea (the more audacious the better), he tried to see if USC and society at large could benefit from it.

Professor Bekey is well known for his own research in robotics, He has had a tremendous impact on the field. He was the founding editor of the *IEEE Journal of Robotics and Automation* (now known as the *IEEE Transactions on Robotics*) and through his own work and his stewardship of the journal shaped the field in its early days. In the mid 1990s he founded and edited a new journal, *Autonomous Robots*, which is published today by Springer. These two journals are among the top 3-4 journals in the field today.

An anecdote comes to mind. In 1986, a young Assistant Professor at MIT, Rodney Brooks, submitted an article to the *IEEE Journal of Robotics and Automation* which presented a radically new way of building controllers for mobile robots. Professor Bekey, as editor, solicited peer reviews of the paper. They all came back recommending that the paper be rejected. In spite of these recommendations, Professor Bekey decided to accept the paper. Hindsight has vindicated his decision. The Brooks paper has since become one of the most cited papers in all of robotics and computer science. Many years later, Professor Bekey was instrumental in hiring one of Professor Brooks' students as a faculty member in robotics at USC - Professor Maja Matarić, currently director of the USC robotics research labs, and senior associate dean in the USC Viterbi School of Engineering.

## 9.5 Recognition, Honors, and Awards

Professor Bekey has been the recipient of many awards over the years. In 1972 he was elected Fellow of IEEE for "Contributions to Hybrid Computation, Man-Machine Systems and Biomedical Engineering". In 1989 he was Elected Member of the National Academy of Engineering for "Pioneering work in Computer Sciences Contributing to Biomedical Engineering, Man-Machine Systems and Robotics". He was elected Fellow of the American Association for Advancement of Science in 1991, and in 1998 he was elected a Fellow of the the American Association for Artificial Intelligence. He was appointed "University Professor" at USC in 1999, and was a recipient of the USC Presidential Medallion in 2000. He is now University Professor Emeritus at USC.

## 9.6 A Personal Tribute

Professor Bekey has had an extraordinary career and an extraordinary life. He insists that he is now retired but his work continues unabated. He has recently published a book, he consults for several companies, and has been inventing new curricula for Computer Science undergraduates. He's full of enthusiasm as always and he's fresh for new challenges even today.

It has been my privilege to know Professor Bekey as a teacher, colleague, mentor, and friend. Its not often one makes the acquaintance of someone who is both inspirational and accessible. I thank him for his help, advice, and encouragement. But mostly I thank him for his willingness - elusive among advisors - to let me pursue my ideas freely.

# Chapter 10
# Current History of the Bekey Tribe

H. Pete Schmid and Monte Ung

The "Bekey Tribe" is comprised of a collection of Professor George Bekey's forty former students, three students of the late legendary professor Richard Bellman, and associates. It exists because we want to maintain a close bond between our Professor and us. The group has met annually since 1984 and the last meeting, the Silver (25th) Anniversary, was held in Arroyo Grande in May 2008. We consider ourselves the living legacy that Professor Bekey presented to society. A montage containing the likenesses of all members is shown on the next page.

Many of the Tribe have joined Academia abroad: Korea University (Gerry Kim '94), National University of Singapore (Huan Liu '89 who has moved back to Arizona State University), Universidad de Santiago (Danilo Bassi '90), Hongkong University of Science & Technology (Dit-Yan Yeung '89). Within the USA our members serve from East, Georgia Tech (Ayanna Howard '99), Yale (John Coggshall '68) to West, UC Davis (Andy Frank '68). Andy Frank is a recognized authority on hybrid electric vehicles and he travels the world promoting this futuristic car and making deals on them. Our members can also be found at the University of Arizona (Tony Lewis '95) and California State Universities (Mohinder Grewal'74 & Tasos Chassiakos '86). In the heartland we are represented at University of Minnesota (Stergios Roumeliotis '00), University of Kansas (Arvin Agah '94), and University of Texas (Gerry Burnham'73), University of New Mexico (Ed Angel '68). Two of us still remain at Troyland (Monte Ung '70 & Gaurav Sukhatme '97). Monte Ung was happily enrolled in the UCLA doctoral program in 1964 when he decided to take a short course with Professor Bekey at USC. At the end of the course he jumped to USC and the rest is history

H. Pete Schmid
Director, Advanced Technology (retd.), Raytheon Missile Systems Company, Tucson, Arizona, e-mail: pete.schmid@alumni.usc.edu

Monte Ung
Department of Electrical Engineering, University of Southern California, Los Angeles, CA 90089, e-mail: ung@ceng.usc.edu

**Fig. 10.1** The Bekey Tribe.

Two of us practice medicine: Louis Nardizzi '67 and Thomas Grove '77 (faculty of Medicine, UCLA). One member answered the call of his faith and became a pastor heading up his own congregation (John Kim '94). Four of us are serving NASA after studying robotics under Professor Bekey. They are Alberto Behar '98, Mike McHenry '98, James Montgomery '99, and Fred Hadaegh '84 who plays an important managerial role in planetary exploration. One of us is a commercial pilot (Jim Maloney '71, flying the Goodyear blimps).

Many among us joined industry such as Northrop-Grumman (Mike Meritt '67), Aerospace Corp (Kirstie Bellman '79), and Orthodyne Inc (Howard Olsen '86). The more adventurous simply formed their own businesses: Simulation Services (Ray Nilsen '68), International Cartoon & Animation (Wissam Ahmed '84) and Research & Development Co. (Dan Antonelli '85).

Some of us, after active careers, have already joined the rank of the "retired": John Coggshall '68, Bayesteh Gaffary '70 actively painting & organizing exhibitions, Justin Biddle '75 back to North Carolina, Pete Schmid '76 actively sailing. He spent forty years in the defense industry - directing advanced weapon technology pursuits for General Dynamics, Hughes and Raytheon.

Four members of the Tribe are deceased: Sam Brainin '66, Anil Phatak '68, Caswell Neal '69, and Jinbom Seo '74.

# Chapter 11
# Recollections and Tributes

Dan Antonelli, Arun Bhadoria, Willis G. Downing, Jr., Huan Liu, Michael Merritt, L. Warren Morrison, and H. Pete Schmid

## 11.1 From Aerospace Engineering to Biomedical Engineering (L. Warren Morrison)

I was a student of Professor Bekey more than forty years ago. After years of Aerospace experience I thought that it was time to return to school and become involved in what we now refer to as Biomedical engineering. Professor Bekey asked me to prepare a research agenda.

Some twenty years later he was cleaning out his files and sent me a letter and enclosed the agenda and a related paper "The N-M-P A Neuro-Muscular Program-

Dan Antonelli
Biokinesiology and Physical Therapy, Biomedical Engineering (retd.), University of Southern California, Los Angeles, CA 90089, e-mail: `dantonel@usc.edu`

Arun Bhadoria
Manufacturing Engineering Leader, Cummins Emission Solutions, Mineral Point, WI 53565, e-mail: `arun.bhadoria@cummins.com`

Willis G. Downing, Jr.
Professor of Biomedical and Electrical Engineering Emeritus, College of Engineering and Computer Science, California State University, Northridge, CA 91330, e-mail: `wgdowning@netptc.net`

Huan Liu
Department of Computer Science and Engineering, Arizona State University, Tempe AZ 85287-8809 e-mail: `huanliu@asu.edu`

Michael Merritt
Mission Systems, Northrup Grumman Corporation e-mail: `michael.merritt@ngc.com`

L. Warren Morrison
Founder and Chief Technology Officer, ITG LABS LLC, e-mail: `warrenmster@gmail.com`

H. Pete Schmid
Director, Advanced Technology (retd.), Raytheon Missile Systems Company, Tucson, Arizona, e-mail: `pete.schmid@alumni.usc.edu`

mer". The work was supported in part under NIH Grant NB 06196-01 - an early example of foresight on the part of NIH and Professor Bekey.

## 11.2 The Final Oral Examination (H. Pete Schmid)

I specifically recall the time of my final oral examination. All the preparation, research, data analysis, and writing of the dissertation were complete - or so I thought.

My committee, headed by Professor George Bekey, had read the fist draft and was well prepared to examine my work and my general preparation for receiving the degree. As we know, this was the final obstacle that had to be overcome. The examination was scheduled at a mutually convenient time. The members of my committee were gathered in the conference room ready for me. As I entered the room I could not have been more nervous. As the examination progressed my nervousness subsided - just a bit. Things went reasonably well, I thought - but I was not too sure.

At the end of the examination I was excused so the committee could discuss my performance and come up with their recommendations. The time I spent waiting outside the conference room seemed to be an eternity. My nervousness returned full force. Finally, after what seemed to be hours, I was invited to return to the conference room to hear the result of the committee's deliberation.

The words I recall are: "You passed, but", my heart sank and I had this terrible knot in my stomach as I listened to the recommendations. In hindsight, it really was not a big deal. The recommendation was for me to combine two chapters covering all the literature search discussion into a single chapter.

After the committee dispersed, Professor. Bekey suggested that I should come to his office. I expected to receive further instructions. But instead, I got a dose of what we all love about Professor Bekey. He said that he knew exactly how I felt, and showed me a picture of a horse that had jumped a fence. The front of the horse was over the fence, but the rear did not make it. That picture and Professor Bekey's understanding of my situation finally put me at ease and lifted the knot off my stomach enough for us to go to lunch.

The moral of the story is that Professor Bekey always has the ability and willingness to provide the required help in any situation. This is just one reason why he has been honored with the large following of the "Bekey Tribe".

## 11.3 Recent Work on Preventing Fractures caused by a Fall (Dan Antonelli)

For several years Professor Bekey has been interested and actively working on a method to prevent hip fractures caused by a fall. Experimental procedures identified the needed sensors and timing to sense the conditions preceding a fall. The problem of an airbag needed some work but we identified the needed sensors and timing to sense the conditions preceding a fall. The algorithm was clear and described the velocity, acceleration and direction of several different falls (forward, backward, sideway or combinations of all three). Empirical tests and computer models confirmed the algorithm and all was set to proceed to a product certain to be a commercial success.

Alas, the holder of the patent refused to listen to sound advice as to how to proceed and the project languishes today, unable to go forward because the patent holder - having only the idea but no clear method to accomplish the goal - has stopped the product from being developed.

## 11.4 Teacher, Mentor, and Friend (Arun Bhadoria)

Professor George Bekey is a very special human being who values feelings and touches hearts. He is the one who supports and helps someone, regardless of their importance, at their time of need -academic or otherwise. It hurts when I remember my early days at USC. While I was pursuing my MS at USC in 2000, after the first semester I had no money for the basic needs of life, and I struggling to get any kind of assistance. Knowledge and prior education were of no use, and many times I stayed awake for more than 3 days without sleeping. Now, as I am remembering those days a few tears are trickling down my cheeks. Along with these tears, a smile starts to play on my lips remembering a person who offered me assistance and held my hand to guide me through that rough time. This was none other than Professor Bekey. It was almost two long years when we worked together every day. Since then he has always stood behind me to help me. Whether it was an issue related to my Indian English, or a severe financial crisis, or offering food knowing that I might be hungry for a few days.

Do I have a close relationship with him? No, I think it can't be defined that simply. Was he my role model? No, as it is not easy to copy him. Is he a hero? Certainly he is and was much more than that. So what is he? If I can define him, he is a 'Guru', and if I have to compare my association with him then he was Lord Krishna of the epic Mahabharata for me. It all started from his big heart. Always smiling, jovial and kind hearted, a very good friend and a teacher who shares both technical knowledge and jokes and laughter.

Every successful man has a woman behind his success. How can I forget Professor Bekey's beautiful wife Shirley at this moment. This note will not be complete if I don't remember all the love and affection showered on my family by Shirley. I remember when I was completely broken and dejected after my wife had her visa rejected. Shirley consoled me saying that far away from home you have us as your family. I still remember those highly soothing and memorable words. There are so many people in this world, but only a few you can count as your real friends and as family.

Now life has given me everything. I have progressed very well professionally in a multi-billion global company and have a beautiful house with a big backyard for my kids in the Midwest. Life always offers new challenges and Professor Bekey's teachings always help me to keep myself motivated and inspired to progress on the right path. Ultimately, as Professor Bekey says, "Play your part and live your life fully and well". And that's what I am trying to do.

I was very fortunate and feel privileged and honored to have had the opportunity to be with Professor Bekey and Shirley. Time can't come back but my family and I feel that time spent with them is our most precious heritage. Salute to my great teacher, my mentor and my friend.

## 11.5  A Testimonial (Willis G. Downing, Jr.)

Like other students of Professor Bekey, I have long admired his many fine qualities. Several incidents during his chairmanship of my dissertation work illustrate some of these qualities, which are: leadership, courageousness, a positive attitude which he transfers to others, a willingness to investigate new ideas, and an ability to organize. I feel that if he had not shared these qualities, the implementation and completion of my dissertation research would have been in doubt.

The first incident happened at a particular time when I felt I had reached an impassible spot in the formulation of a research protocol. Professor Bekey's response was (paraphrased), "you can solve it and you can solve it by X a.m. tomorrow." Oddly enough, I worked it out by that time. Professor Bekey also suggested the original research question, which was challenging to accepted concepts. Also, illustrating more courageousness, he participated as a subject in my experimentation, which used human subjects.

For all of this, and much else, I am grateful to Professor Bekey.

## 11.6  Making it Look Easy (Huan Liu)

When I joined Professor Bekey's research group in the beginning of 1986, he also served as the department chair. I was quite nervous whether he would even find time to talk me as he already had some very capable students in his lab. I was so relieved and happy that he kindly agreed to be my thesis advisor after a pleasant meeting. From then on, Professor Bekey found time in his busy schedule to patiently guide me into research as I had very little experience in research. I often told my students how busy my advisor was as I vividly remembered that our regularly research meetings were often interrupted. What amazed me is his superb multi-tasking capability: it looked so easy and natural for him to effortlessly move from one task to another. In addition to his patience, he truly trusted me and offered me the great opportunity to work on the dextrous robot hand project. When I moved to Arizona State University from the National University of Singapore in the begining of 2000, Professor Bekey again provided tremendous support and invaluable advice as everything was so strange and new to me after spending 10 years in Australia and Singapore. I genuinely hope I could be as patient and supportive to my students as he was and is toward me. When we meet next time, I will ask Professor Bekey the following questions: (1) while remaining patient and supportive, how he could still be so productive in getting students out of USC, and (2) what are the tricks for his effortless multi-tasking that I really need to learn. After becoming a professor, I become more and more grateful for what Professor Bekey has done for me, and I am trying my best to graduate quality students who can make their great contributions to our fast changing field.

## 11.7 Solving Complex Problems Efficiently (Michael Merritt)

Professor Bekey founded the Analog Hybrid Lab at USC (housed in Olin Hall). The digital computer was originally an IBM 1710. This was really an IBM 1620 with additional features for analog computation: A/D converters, D/A converters, general-purpose I/O lines, and interrupts. It was small and slow and had a lot of quirks (like divide overflow and stop). Professor Bekey obtained funding to replace it with an IBM Mainframe - an ambitious project at the time. The IBM 360 computers were new and in great demand and any delays in getting the order placed would substantially delay the delivery. At lunch with the IBM representative Professor Bekey wrote the letter of intent for the computer on a napkin and signed it. This insured that USC had a position in the assembly line and that, pending a more formal order, it would be delivered quickly. While the words IBM Mainframe have lost much of their meaning in the intervening years, at the time it represented a large room full of computer equipment and a substantial monetary investment. Further, IBM at the time was a very formal and straitlaced company. The IBM representative was a bit startled, as you can imagine, but it is an example of Professor Bekey's ability to solve complex problems in a direct and efficient way.

# References

[1] Moore's law. http://en.wikipedia.com/wiki/Moore's\_law

[2] Nvida cuda compute unified device architecture programming manu al. http://nvidia.com/cuda

[3] 20th century fox: Star wars (the movie). en.wikipedia.org/wiki/Star_Wars (1977)

[4] A. Guccione, e.a.: Development and testing of a self-report instrument to measure actions: Outpatient physical therapy improvement in movement assessment log (optimal). Physical Therapy **85**, 515–530 (2005)

[5] Açı kmeşe, B., D.P., S., Carson, J., Hadaegh, F.: Distributed estimation for spacecraft formationsover time-varying sensing topologies. In: Proceedings of the 17th IFAC World Congress, pp. 2123–2130 (2008)

[6] Açı kmeşe, B., Scharf, D., Murray, E., Hadaegh, F.: A convex guidance algorithm for formation reconfiguration. In: AIAA Guidance, Navigation and Control Conference (2006)

[7] Acikmese, B., Hadaegh, F., Scharf, D., S.R., P.: Formulation and analysis of stability for spacecraft formations. IET Control Theory and Application, special issue on Cooperative Control of Multiple Spacecraft Flying in Formation **1**(2), 461–474 (2007)

[8] Agah, A., Bekey, G.A.: Phylogenetic and Ontogenetic Learning in a Colony of Interacting Robots. Autonomous Robots **4**(1), 85–100 (1997)

[9] Akers, E.L., Agah, A.: Design and Simulation of a Polar Mobile Robot. Journal of Intelligent Systems **17**(4), 379–404 (2008)

[10] Akers, E.L., Harmon, H.P., Stansbury, R.S., Agah, A.: Design, Fabrication, and Evaluation of a Mobile Robot for Polar Environments. In: IEEE International Geoscience and Remote Sensing Symposium (IGARSS), pp. 109–112. Proceedings of the IEEE International Geoscience and Remote Sensing Symposium, Anchorage, Alaska (2004)

[11] Akers, E.L., Stansbury, R.S., Agah, A.: Long-term Survival of Polar Mobile Robots. In: Proceedings of the 4th International Conference on Computing, Communications and Control Technologies (CCCT), vol. II, pp. 329–333. Orlando, FL (2006)

[12] Akers, E.L., Stansbury, R.S., Agah, A., Akins, T.L.: Mobile Robots for Harsh Environments: Lessons Learned from Field Experiments. In: Proceedings of the 11th International Symposium on Robotics and Applications (ISORA), pp. 1–6. Budapest, Hungary (2006)

[13] Ambrose, R.O., Aldridge, H., Askew, R.S., Burridge, R.R., Bluethmann, W., Diftler, M., Lovchik, C., Magruder, D., Rehnmark, F.: Robonaut: Nasa's space humanoid. IEEE Intelligent Systems **15**(4), 57–63 (2000)

[14] Arbib, M.: The Metaphorical Brain 2: Neural Networks and Beyond. Wiley-Interscience, New York (1989)

[15] Arcone, S.: Personal Communications. Cold Regions Research and Engineering Lab (CRREL), Hanover, New Hampshire (2003)

[16] Arkin, R.: Behavior-Based Robotics. The MIT Press, Cambridge (1998)

[17] Asada, H., Slotine, J.: Robot Analysis and Control. John Wiley and Sons, New York (1986)

[18] B., S., Plaisant, C.: Designing the User Interface. Addison Wesley (2004)

[19] Baxter, J.: A Bayesian/information theoretic model of bias learning. In: Proceedings of the Ninth Annual Conference on Computational Learning Theory, pp. 77–88. Desenzano del Garda, Italy (1996)

[20] Baxter, J.: A Bayesian/information theoretic model of learning to learn via multiple task sampling. Machine Learning **28**(1), 7–39 (1997)

[21] Beard, R., McLain, T., Hadaegh, F.: Fuel optimization for constrained rotation of spacecraft formations. Journal of Guidance, Control and Dynamics **23**(2), 339–346 (2001)

[22] Beer, R., Chiel, H.J., Quinn, R., Ritzmann, R.: Biorobotic approaches to the study of motor systems. Current Opinion in Neurobiology **8**, 777–782 (1998)

[23] Beer, R., R.D., Q., Chiel, H.J., Ritzmann, R.: Biologically-inspired approaches to robotics. Communication of the ACM **40**(3) (1997)

[24] Beer, R.D., Chiel, H.J., Gallagher, J.: Evolution and analysis of model cpgs for walking ii. general princ iples and individual variability. Journal of Computational Neuroscience **7**, 119–147 (1999)

[25] Beer, R.D., Chiel, H.J., Gallagher, J.C.: Evolution and analysis of model cpgs for walking i. dynamic modules' Journal of Computational Neuroscience **7**, 99–118 (1999)

[26] Bekey, G.: Autonomous Robots. MIT Press, Cambridge, Massachusetts (2005)

[27] Bekey, G.: Autonomous Robots: From Biological Inspiration to Implementation a nd Control. Blackwell Scientific Publications, The MIT PRESS (2005)

[28] Bekey, G.A., Agah, A.: Group Behavior of Robots, second edn., pp. 439–445. Shimon Y. Nof (Ed.) Handbook of Industrial Robotics. John Wiley & Sons Inc., New York, New York (1999)

[29] Biernacki, C., Celeux, G., Govaert, G.: Assessing a mixture model for clustering with the integrated completed likelihood. IEEE Transactions on Pattern Analysis and Machine Intelligence **22**(7), 719–725 (2000)

[30] Bingham, C., Mardia, K.V.: A small circle distribution on the sphere. Biometrika **65**(2), 379–389 (1978)

[31] Blum, A., Chawla, S., Karger, D.R., Lane, T., Meyerson, A., Minkoff, M.: Approximation algorithms for orienteering and discounted-reward tsp. In: Proceedings of the 44th Annual IEEE Symposium on Foundations of Computer Science, pp. 46–55 (2003)

[32] Bonilla, E., Agakov, F., Williams, C.: Kernel multi-task learning using task-specific features. In: Proceedings of the Eleventh International Conference on Artificial Intelligence and Statistics. San Juan, Puerto Rico (2007)

[33] Bonilla, E., Chai, K., Williams, C.: Multi-task Gaussian process prediction. In: J. Platt, D. Koller, Y. Singer, S. Roweis (eds.) Advances in Neural Information Processing Systems 20, pp. 153–160. MIT Press, Cambridge, MA, USA (2008)

[34] Brock, O., Fagg, A.H., Grupen, R.A., Karuppiah, D., Platt, R., Rosenstein, M.: A framework for humanoid control and intelligence. International Journal of Humanoid Robotics **2**(3), 301–336 (2005)

[35] Brooks, R.: Achieving artificial intelligence through building robots. Tech. rep., Massachusetts Institute of Technology (1986)

[36] Brooks, R., Stein, A.: Building brains for bodies. Autonomous Robots **1**, 7–25 (1994)

[37] Brown, T.G.: The intrinsic factors in the act of progression in the mammal. Proc. of the Royal Soc. Lond., Ser. B. **84**, 309–319 (1911)

[38] Burgard, W., Moors, M., Stachniss, C., Schneider, F.E.: Coordinated multi-robot exploration. IEEE Transactions on Robotics **21**(3) (2005)

[39] Burridge, R.R., Graham, J., Shillcutt, K., Hirsh, R., Kortenkamp, D.: Experiments with an EVA Assistant Robot. In: Proceedings of the 7th International Symposium on Artificial Intelligence, Robotics and Automation in Space (I-SAIRAS-03) (2003)

[40] Canny, J., Reif, J., Donald, B., Xavier, P.: On the complexity of kinodynamic planning. In: 29th IEEE Annual Symp. on Foundations of Computer Science, pp. 306–316 (1988)

[41] Carmichael, B.L., Gifford, C.M.: Modeling and Simulation of the Seismic TETwalker Concept. Tech. Rep. CReSIS-TR-134 (2007)

[42] Chaudhuri, K., Godfrey, B., Rao, S., Talwar, K.: Paths, trees and minimum latency tours. In: Proceedings of the 44th Annual IEEE Symposium on Foundations of Computer Science, pp. 36–45 (2003)

[43] Ciocarlie, M., Goldfeder, C., Allen, P.: Dexterous grasping via eigengrasps: A low-dimensional approach to a high-complexity problem. In: Proceedings of the Robotics: Science & Systems 2007 Workshop - Sensing and Adapting to the Real World (2007). Electronically published

[44] Coelho, Jr., J.A., Grupen, R.A.: A control basis for learning multifingered grasps. Journal of Robotic Systems **14**(7), 545–557 (1997)

[45] Coelho, Jr., J.A., Piater, J., Grupen, R.A.: Developing haptic and visual perceptual categories for reaching and grasping with a humanoid robot. Robotics and Autonomous Systems Journal, special issue on Humanoid Robots **37**(2–3), 195–219 (2000)

[46] Craig, J.: Introduction to Robotics: Mechanics and Control, third edn. Prentice Hall (2004)

[47] Craig, J.J.: Introduction to Robotics: Mechanics and Control. Printice Hall (2004)

[48] CReSIS: Center for Remote Sensing of Ice Sheets. URL: http://www.cresis.ku.edu/ (2006)

[49] Curtis, S., Mica, J., Nuth, J., Marr, G.: ANTS (Autonomous Nano Technology Swarm): An Artificial Intelligence Approach to Asteroid Belt Resource Exploration. In: Proceedings of the 51st International Astronautical Congress. Rio de Janeiro, Brazil (2000)

[50] C.Walshaw, Cross, M.: Mesh partitioning: A multilevel balancing and refinement algorithm. SIAM Journal on Scientific Computing $22(1)$, 63–80 (2000)

[51] Cyberbotics: Webots 5. URL: http://www.cyberbotics.com/ (2006)

[52] Dautenhahn, K., Billard, A.: Games children with autism can play with robota, a humanoid robotic doll. In: Proceedings of Cambridge Workshop on Universal Access and Assistive Technology, pp. 179–190 (2002)

[53] Dayan, P., Abbott, L.F.: Theoretical Neuroscience: Computational and mathematical Modeling o f Neural Systems. The MIT Press, Cambridge, MA (2001)

[54] de Granville, C., Fagg, A.H.: Learning grasp affordances through human demonstration (submitted)

[55] de Granville, C., Southerland, J., Fagg, A.H.: Learning grasp affordances through human demonstration. In: Proceedings of the International Conference on Development and Learning (2006). Electronically published

[56] Delcomyn, F.: Neural basis of rhythmic behavior in animals. Science $210$, 492–498 (1980)

[57] Dempster, A., Laird, N., Rubin, D.: Maximum likelihood estimation from incomplete data via the EM algorithm. Journal of the Royal Statistical Society, Series B $39(1)$, 1–38 (1977)

[58] Deo, N.: Graph Theory with Applications to Engineering and Computer Science. Prentice-Hall (1974)

[59] Dhariwal, A., Zhang, B., Oberg, C., Stauffer, B., Requicha, A., Caron, D., Sukhatme, G.S.: Networked aquatic microbial observing system. In: the Proceedings of the IEEE International Conference of Robotics and Automation (ICRA) (2006)

[60] Eiken, O., Degutsch, M., Riste, P., Rod, K.: Snowstreamer: An Efficient Tool in Seismic Acquisition. First Break $7(9)$, 374–378 (1989)

[61] Ekeberg O Grillner, S.: Simulations of neuromuscular control in lamprey swimming. Philos Trans R Soc Lond B Biol Sci. $354(1385)$ (1999)

[62] Ellis, H.: The Transfer of Learning. Macmillan, New York (1965)

[63] Escobedo, T.H.: Play at the art table: A study of children's play behaviors while drawing. In: Annual Conference of the Association for the Study of Play (1996)

[64] Fagg, A.H., Rosenstein, M.T., Platt, Jr., R., Grupen, R.A.: Extracting user intent in mixed initiative teleoperator control. In: Proceedings of the American Institute of Aeronautics and Astronautics Intelligent Systems Technical Conference (2004). Electronically published

[65] Fox, D., Ko, J., Konolige, K., Limketkai, B., Schulz, D., Stewart, B.: Distributed multirobot exploration and mapping. IEEE Proceeding **94**(7), 1325–1339 (2006)

[66] Galloway, J., Rhu, J., Agrawal, S.: Babies driving robots: Self-generated mobility in very young infants. Journal of Intelligent Service Robotics (2008)

[67] Garmin Ltd.: GPSMAP 76S. URL: http://www.garmin.com/products/gpsmap76s (2005)

[68] Gates, W.: A robot in every home. Scientific American (2007)

[69] Gibson, J.J.: The Senses Considered as Perceptual Systems. Allen and Unwin (1966)

[70] Gibson, J.J.: The theory of affordances. In: R.E. Shaw, J. Bransford (eds.) Perceiving, Acting, and Knowing. Lawrence Erlbaum, Hillsdale (1977)

[71] Gifford, C.M.: Robotic Seismic Sensors for Polar Environments. Master's Thesis, Department of Electrical Engineering and Computer Science, University of Kansas (2006)

[72] Gifford, C.M., Agah, A.: Robotic Deployment and Retrieval of Seismic Sensors for Polar Environments. In: Proceedings of the 4th International Conference on Computing, Communications and Control Technologies (CCCT), vol. II, pp. 334–339. Orlando, FL (2006)

[73] Gifford, C.M., Agah, A.: Precise Formation of Multi-Robot Systems. In: Proceedings of the IEEE International Conference on Systems of Systems Engineering (SoSE), pp. 1–6. San Antonio, TX (2007)

[74] Gifford, C.M., Agah, A., Tsoflias, G.P.: Hybrid Streamers for Polar Seismic. Eos Trans. AGU **87**(52) (2006)

[75] Ginsburg, K.R.: The importance of play in promoting healthy child development and maintaining strong parent-child bonds (2006)

[76] Ginsburg, K.R.: The importance of play in promoting healthy child development and maintaining strong parent-child bonds. Pediatrics **119** (2007)

[77] Gockley, R., Mataric, M.: Encouraging physical therapy compliance with a hands-off mobile robot. In: Proc. of Human-Robot Interaction (2006)

[78] de Granville, C.: Learning grasp affordances. Master's thesis, School of Computer Science, University of Oklahoma, Norman, OK (2008)

[79] Grillner, S., Zangger, P.: On the central generation of locomotion in the low spinal cat. Exp. Brain Res. **34**, 241–61 (1979)

[80] Hand, D.J., Smyth, P., Mannila, H.: Principles of Data Mining. MIT Press, Cambridge, MA, USA (2001)

[81] Harmon, H.P., Stansbury, R.S., Akers, E.L., Agah, A.: Sensing and Actuation for a Polar Mobile Robot. In: Proceedings of the International Conference on Computing, Communications and Control Technologies (CCCT), vol. IV, pp. 371–376 (2004)

[82] Hille, B.: Ionic Channels of Excitable Membranes. Sinauer, Sunderland (1984)

[83] Hodgkin, A.L., Huxley, A.F.: A quantitative description of membrane current and its application to conduction and excitation in nerve. J. Physiol. **117**(4), 500–544 (1952)

[84] Hollis, J., Iseli, J., Williams, M., Hoenmans, S.: The Future of Land Seismic. Hart's E & P **78**(11), 77–81 (2005)

[85] Hornik, K., White, M.S.H.: Multilayer feedforward networks are universal approximators. Neural Networks **2**, 359–366 (1989)

[86] Howard, A., Bekey, G.: Intelligent learning for deformable object manipulation. Autonomous Robots **9**, 51–58 (2000)

[87] Howard, A., Park, H., Kemp, C.: Extracting play primitives for a robot playmate by sequencing low-level motion behavior. In: Int. Symp. on Robot and Human Interactive Communication (RO-MAN). IEEE, Munich, Germany (2008)

[88] Howard, A., Remy, S., Park, H.: Learning of arm exercise behaviors: Assistive therapy based on therapist-patient observation. In: RSS: Workshop on Interactive Robot Learning. Zurich, Switzerland (2008)

[89] Hussein, I., Schaub, H.: Invariant shape solutions of the spinning three craft coulomb tether problem. In: Advances in the Astronautical Sciences, Spaceflight Mechanics, Proceedings of the AAS/AIAA Space Flight Mechnaics Meeting, vol. 124, pp. 2015–2033 (2006)

[90] Izhikevich, E.: Simple model of spiking neurons. IEEE Transactions on Neural Networks **14**(6), 1569 – 1572 (2003)

[91] Izhikevich, E.: Which model to use for cortical spiking neurons? IEEE Transactions on Neural Networks **15**(5), 1063 – 1070 (2004)

[92] Kansas Geological Survey: LandStreamer. URL: `http://www.kgs.ku.edu/Geophysics2/Equip/LandStreamer/LandS4.htm` (2006)

[93] Kimura, H., Fukuoka, Y., Hada, Y., Takase, K.: Adaptive dynamic walking of a quadruped robot on irregular terrain using a neural system model. In: ISRR, pp. 88–97. Lorne, Australia (2001)

[94] Kimura, H., Fukuoka, Y., Konaga, K., Hada, Y., Takase, .K.: Towards 3d adaptive dynamic walking of a quadruped robot on irregul ar terrain by using neural system model. In: IEEE and RSJ IROS 2001. IEEE, Hawaii (2001)

[95] King, E.C., Bell, A.C.: A Towed Geophone System for use in Snow-Covered Terrain. Geophysical Journal International **126**(1), 54–62 (1996)

[96] King, L., Parker, G., Deshmukh, S., Chong, J.H.: Study of interspacecraft coulomb forces and implications for formation flying. Journal of Propulsion and Power **19**(3), 497–505 (2003)

[97] Kong, E., Kwon, D., Schweighart, S., Elias, L., Sedwick, R., Miller, D.: Electromagnetic formation flight for multisatellite arrays. Journal of Guidance, Control and Dynamics **41**(4), 659–666 (2005)

[98] Krause, A., Guestrin, C., Gupta, A., Kleinberg, J.: Near-optimal sensor placements: Maximizing information while minimizing communiction cost. In:

Proccedings of the Fifth International Conference on Information Processing in Sensor Networks (IPSN'06), pp. 2–10 (2006)

[99] Kronreif, G., Prazak, B., Mina, S., Kornfeld, M., Meindl, M., Furst, F.: Games children with autism can play with robota, a humanoid robotic doll. In: Proceedings of Cambridge Workshop on Universal Access and Assistive Technology, pp. 179–190 (2002)

[100] KU PRISM Team: PRISM Home: Polar Radar for Ice Sheet Measurements. URL: http://www.ku-prism.org (2004)

[101] Kurz, D., Hantsch, F., Grosse, M., Schiewe, A., Bimber, O.: Laser pointer tracking in projector-augmented architectural environments. In: Proc. of ISMAR, pp. 19–26 (2007)

[102] Lawrence, N., Platt, J.: Learning to learn with the informative vector machine. In: Proceedings of the Twenty-First International Conference on Machine Learning, pp. 65–72. Banff, Alberta, Canada (2004)

[103] Lawson, P.R.: The terrestrial planet finder. International Aeronautical Congress pp. 2005–2011 (2001)

[104] Lewis, M.A.: Perception driven robot locomotion. Robot Society of Japan Journal. (2002)

[105] Lewis, M.A., Etienne-Cummings, R., Hartmann, M., Cohen, .A.: Toward biomorphic control using custom avlsi chips. In: 2000 International Conference on Robotics and Automation. IEEE, San Francisco (2000)

[106] Lewis, M.A., Etienne-Cummings, R., Hartmann, M., Xu, Z.R., Cohen, A.H.: An in silico central pattern generator: Oscillator, entrainment, m otor neuron adaptation & biped mechanism control. Biological Cybernetics **88**(2), 137–151 (2003)

[107] Lewis, M.A., Nelson, M.E.: Look before you leap: Peering behavior for depth perception. In: Simulation of Adaptive Behavior. Zurich (1998)

[108] Lindblom, J., Ziemke, T.: Analysis of the back-propagation algorithm with momentum. IEEE Transactions on Neural Networks **5**, 505–506 (1994)

[109] Lowe, D.G.: Distinctive image features from scale-invariant keypoints. International Journal of Computer Vision **60**(2), 91–110 (2004)

[110] Mardia, K.V., Jupp, P.E.: Directional Statistics. Wiley Series in Probability and Statistics. Wiley, Chichester, West Sussex, England (1999)

[111] Matsuoka, K.: Mechanisms of frequency and pattern control in the neural rhythm ge nerators. Biol. Cybern **56**, 345–353 (1987)

[112] Metchev, S., Grindlay, J.: A two-dimensional kolmogorov-smirnov test for crowded field source detection: Rosat sources in ngc 6397. In: Mon Not R Astron Soc., v335, pp. 73–83 (2002)

[113] Metrica, Inc.: Mars Manipulator. URL: http://www.metricanet.com/mars.htm (2006)

[114] Michaud, F.: Assistive technologies and child-robot interaction. In: Proceedings of American Association for Artificial Intelligence Spring Symposium on Multidisciplinary Collaboration for Socially Assistive Robotics (2007)

[115] Minka, T., Picard, R.: Learning how to learn is learning with point sets. Tech. rep., MIT Media Lab (1999)

[116] MSC Software: visualNastran 4D R2 User Manual (2002)

[117] Müller, W.G.: Collecting Spatial Data. Physica-Verlag (2001)

[118] National Science Foundation: Amundsen-Scott South Pole Station. URL: http://www.nsf.gov/od/opp/support/southp.jsp (2006)

[119] Nguyen-Tuong, D., Peters, J., Seeger, M., Schölkopf, B.: Learning inverse dynamics: a comparison. In: Proceedings of the European Symposium on Artificial Neural Networks. Bruges, Belgium (2008)

[120] Nguyen-Tuong, D., Seeger, M., Peters, J.: Computed torque control with nonparametric regression models. In: Proceedings of the 2008 American Control Conference. Seattle, Washington, USA (2008)

[121] Niels Bohr Institute: NGRIP: North Greenland ice core project. URL: http://www.glaciology.gfy.ku.dk/ngrip/index_eng.htm (2005)

[122] Nourbakhsh, I.: Course notes for principles of human robot interaction (2007). (Unpublished)

[123] Park, C., Howard, A.: Haptically guided teleoperation for learning manipulation tasks. In: Robotics: Science and Systems: Workshop on Robot Manipulation. Atlanta, GA (2007)

[124] Park, C., Howard, A.: Vision-based force guidance for improved human performance in a teleoperative manipulation system. In: Int. Conf. on Intelligent Robots and Systems (IROS). IEEE/RSJ, San Diego, CA (2007)

[125] Paul, R.: Robot Manipulators. MIT Press, Cambridge, Massachusetts, USA (1981)

[126] Paul, R.P.: Robot Manipulators: Mathematics, Programming, and Control. The MIT Press, Cambridge (1982)

[127] Phansalkar, V.V., Sastry, P.S.: Analysis of the back-propagation algorithm with momentum. IEEE Transactions on Neural Networks **5**, 505–506 (1994)

[128] Piaget, J.: Play, Dreams and Imitation in Childhood. Routledge and Kegan Paul Ltd, London, UK (1951)

[129] Piater, J.H., Grupen, R.A.: Feature learning for recognition with Bayesian networks. In: Proceedings of the Fifteenth International Conference on Pattern Recognition. Barcelona, Spain (2000)

[130] Piater, J.H., Grupen, R.A.: Learning appearance features to support robotic manipulation. In: Proceedings of the Cognitive Vision Workshop (2002). Electronically published

[131] Pinhanez, C.: Creating ubiquitous interactive games using everywhere display projectors. In: Proc. of IWEC

[132] Platt, Jr., R.: Learning and generalizing control-based grasping and manipulation skills. Ph.D. thesis, Department of Computer Science, University of Massachusetts, Amherst (2006)

[133] Platt, Jr., R., Fagg, A.H., Grupen, R.A.: Nullspace composition of control laws for grasping. In: Proceedings of the International Conference on Intelligent Robots and Systems, pp. 1717–1723 (2002)

[134] Platt, Jr., R., Fagg, A.H., Grupen, R.A.: Whole body grasping. In: Proceedings of International Conference on Robotics and Automation (ICRA'03) (2003)

[135] Ploen, S., Scharf, D., Hadaegh, F.: Dynamics and control of drag-free formations for earth imaging applications. In: SPIE International Asia-Pacific Symposium: Remote Sensing of the Atmosphere, Environment and Space

[136] R. E. Sheriff and L. P. Geldart: Exploration Seismology, second edn. Cambridge University Press (1995)

[137] Rahimi, M.H.: Bioscope: Actuated sensor network for biological science. Ph.D. thesis, Computer Science Department, Viterbi School of Engineering, USC (2005)

[138] Rancourt, D., Rivest, L.P., Asselin, J.: Using orientation statistics to investigate variations in human kinematics. Applied Statistics **49**(1), 81–94 (2000)

[139] Raskar, R., Cutts, M., Welch, G., Stuerzlinger, W.: Efficient image generation for multiprojector and multisurface displays. In: Ninth EuroGraphics Rendering Workshop (1998)

[140] Rasmussen, C., Williams, C.: Gaussian Processes for Machine Learning. MIT Press, Cambridge, Massachusetts, USA (2006)

[141] Recreative Industries: Buffalo All Terrain Truck. URL: http://www.maxatvs.com/ (2004)

[142] Remy, S., Howard, A.: In situ interactive teaching of trustworthy robotic assistants. In: IEEE Int. Conf. on Systems, Man, and Cybernetics (2007)

[143] Ren, W., Beard, R.W.: Decentralized scheme for spacecraft formation flying via the virtual structure approach. Journal of Guidance, Control and Dynamics **27**(1), 73–82 (2004)

[144] Rivest, L.P.: A directional model fo the statistical analysis of movement in three dimensions. Biometrika **88**(3), 779–791 (2001)

[145] Ruiz-Carrion, I., Gifford, C.M.: Detailed Modeling of Designs for the Polar Seismic TETwalker. Tech. Rep. CReSIS-TR-132 (2007)

[146] Ruppert, D., Wand, M.P.: Multivariate locally weighted least squares regression. The Annals of Statistics **22**(3), 1346–1370 (1994)

[147] Sandhu, J., Mesbahi, M., Tsukamaki, T.: Relative sensing networks: Observability, estimation, and the control structure. In: Proceedings of the IEEE Conference on Decision and Control, pp. 6400–6405 (2005)

[148] Santello, M., Flanders, M., Soechting, J.F.: Postural hand synergies for tool use. Journal of Neuroscience **18**(23), 10,105–10,115 (1998)

[149] Scassellati, B.: How social robots will help us to diagnose, treat, and understand autism. In: 12th International Symposium of Robotics Research (2005)

[150] Scharf, D., Acikmese, B., Ploen, S., Hadaegh, F.: A direct solution for fuel-optimal reactive collision avoidance of collaborating spacecraft. In: American Control Conference, pp. 5201–5206 (2006)

[151] Scharf, D., Hadaegh, F., Ploen, S.: A survey of spacecraft formation flying guidance and control (part i): Guidance. In: American Control Conference (2003)

[152] Scharf, D., Hadaegh, F., Ploen, S.: A survey of spacecraft formation flying guidance and control (part ii): Control. In: American Control Conference (2004)

[153] Schölkopf, B., Smola, A.: Learning with Kernels. MIT Press, Cambridge, MA, USA (2002)

[154] Schwaighofer, A., Tresp, V., Yu, K.: Learning Gaussian process kernels via hierarchical Bayes. In: L. Saul, Y. Weiss, L. Bottou (eds.) Advances in Neural Information Processing Systems 17, pp. 1209–1216. MIT Press, Cambridge, MA, USA (2005)

[155] Sen, V., Stoffa, P.L., Dalziel, I.W.D., Blankenship, D.D., Smith, A.M., Anandakrishnan, S.: Seismic Surveys in Central West Antarctica: Data Processing Examples from the ANTALITH Field Tests. Terra Antarctica 5(4), 761–772 (1999)

[156] Sethares, W.A., Staley, T.W.: Periodicity transforms. In: IEEE Trans. on Signal Processing, pp. 2953–2964 (1999)

[157] Simmons, R.G., Apfelbaum, D., Burgard, W., Fox, D., Moors, M., Thrun, S., Younes, H.: Coordination for multi-robot exploration and mapping. In: AAAI, pp. 852–858 (2000)

[158] Singh, A., Krause, A., Guestrin, C., Kaiser, W., Batalin, M.: Efficient planning of informative paths for multiple robots. In: Proceedings of the International Joint Conference on Artificial Intelligence, pp. 2204–2211 (2007)

[159] Singh, G., Hadaegh, F.: Collision avoidance guidance for formation-flying applications. In: AIAA Guidance, Navigation and Control Conference (2001)

[160] Smith, R., F.Y., H.: A distributed parallel estimation architecture for cooperative vehicle control. In: American Control Conference, pp. 4219–4224 (2006)

[161] Smith, R., Hadaegh, F.: Control of deep space formation flying spacecraft: Relative sensing and switched information. AIAA Journal of Guidance, Control and Dynamics (2005)

[162] Smith, R., Hadaegh, F.: Closed-loop dynamics of cooperative vehicle formations with parallel estimators and communication. IEEE Transactions on Automatic Control 52(8), 1404–1414 (2007)

[163] Smith, R.S., Y., H.F.: Distributed estimation, communication and control for deep space formations. IET Control Theory and Application, special issue on Cooperative Control of Multiple Spacecraft Flying in Formation 1(2), 445–451 (2007)

[164] Speece, M.A., Miller, C.R., Miller, P.F., Link, C.A., Flynn, K.F., Dolena, T.M.: A Rapid-Deployment, Three Dimensional (3-D), Seismic Reflection System. Montana Tech. Prototype Design Proposal (2004)

[165] Spikes, K., Steeples, D., Ralston, M., Blair, J., Tian, G.: Common Midpoint Seismic Reflection Data Recorded with Automatically Planted Geophones. URL: http://www.dot.ca.gov/hq/esc/geotech/gg/geophysics2002/037spikes_cmp_auto_plants.pdf (2001)

[166] Stansbury, R.S., Akers, E.L., Harmon, H.P., Agah, A.: Survivability, Mobility, and Functionality of a Rover for Radars in Polar Regions. International Journal of Control, Automation, and Systems 2(3), 334–353 (2004)

[167] Stroupe, A.W., Ravichandran, R., Balch, T.: Value-based action selection for exploration and dynamic target observation with robot teams. In: Proceedings

of the IEEE International Conference on Robotics and Automation, pp. 4190–4197 (2004)

[168] Sukhatme, G.S., Dhariwal, A., Zhang, B., Oberg, C., Stauffer, B., Caron, D.A.: The design and development of a wireless robotic networked aquatic microbial observing system. Environmental Engineering Science **24**(2), 205–215 (2006)

[169] Swaroop, D., Hedrick, J.: String stability of interconnected systems. IEEE Transactions on Automatic Control **41**(3), 349–357 (1996)

[170] Taga, G.: A model of the neuro-musculo-skeletal system for human locomotion. i. emergence of basic gait. Biological Cybernetics **73**, 97–111 (1995)

[171] Taga, G.: A model of the neuro-musculo-skeletal system for human locomotion. ii. real-time adaptability under various constraints. Biological Cybernetics **73**, 113–121 (1995)

[172] Taga, G.: A model of the neuro-musculo-skeletal system for anticipatory adjustment of human locomotion during obstacle avoidance. Biological Cybernetics **78**, 9–17 (1998)

[173] Taga, G., Yamaguchi, Y., Shimizu, H.: Self-organized control of bipedal locomotion by neural oscillators in unpredictable environment. Biol. Cybern. **65**, 147–159 (1991)

[174] TerraTrack Inc.: RangeRunner by TerraTrack. URL: http://www.terratrack.com (2005)

[175] Thrun, S.: Is learning the $n$-th thing any easier than learning the first? In: D. Touretzky, M. Mozer, M. Hasselmo (eds.) Advances in Neural Information Processing Systems 8, pp. 640–646. MIT Press (1996)

[176] Thrun, S., Burgard, W., Fox, D. (eds.): Probabilistic Robotics. MIT Press (2005)

[177] Tien, J., Purcell, G., Amaro, L., Young, L., M. Aung, M., Srinivasan, J., Archer, E., Vozoff, A., Chong, Y.: Technology validation of the autonomous formation flying sensor for precision formation flying. Aerospace Conference Proceedings, IEEE **1**, 1–140 (2003)

[178] Tokuda, Y., Iwasaki, S., Sato, Y., Nakanishi, Y., Koike, H.. Ubiquitous display for dynamically changing environments. In: Proc. of CHI

[179] Topcon Positioning Systems, Inc.: Legacy GPS+ Receiver. URL: http://www.topconeurope.com/index.asp?pageid=12166e1461284b7ca5e68c4f0257448d (2008)

[180] Tsoflias, G.P., Steeples, D.W., Czarnecki, G.P., Sloan, S.D., Eslick, R.C.: Automatic Deployment of a 2-D Geophone Array for Efficient Ultra-Shallow Seismic Imaging. Geophysical Research Letters (2006)

[181] Van der Veen, M., Green, A.: Land Streamer for Shallow Seismic Data Acquisition: Evaluation of Gimbal-Mounted Geophones. Geophysics **63**, 1408–1413 (1998)

[182] Van der Veen, M., Wild, P., Spitzer, R., Green, A.: Design Characteristics of a Seismic Land Streamer for Shallow Data Acquisition. In: Extended Abstracts of the 61st European Association of Geoscientists and Engineers Conference and Technical Exhibition, pp. 40–41 (1999)

[183] Vapnik, V.: Statistical Learning Theory. Wiley, New York, NY, USA (1998)

[184] Vidyasagar, M.: Nonlinear System Analysis, 2nd edn. Prentice-Hall, Inc., Englewood Cliffs, NJ (1993)

[185] Vijayakumar, S., D'Souza, A., Schaal, S.: Incremental online learning in high dimensions. Neural Computation **17**(12), 2602–2634 (2005)

[186] Vijayakumar, S., D'Souza, A., Shibata, T., Conradt, J., Schaal, S.: Statistical learning for humanoid robots. Autonomous Robots **12**(1), 55–69 (2002)

[187] Vijayakumar, S., Schaal, S.: Locally weighted projection regression: an $o(n)$ algorithm for incremental real time learning in high dimensional space. In: Proceedings of the Seventeenth International Conference on Machine Learning, pp. 1079–1086. Stanford, CA, USA (2000)

[188] Vogt, F., Wong, J., Fels, S., Cavens, D.: Tracking multiple laser pointers for large screen interaction. In: Proc. of UIST (2003)

[189] Wade, U.B., Gifford, C.M.: Investigation of Power Sources for the Polar Seismic TETwalker. Tech. Rep. CReSIS-TR-133 (2007)

[190] Wakiji, E.: Mapping the literature of physical therapy. Bull Med Libr Assoc. **85**, 284–288 (1997)

[191] Wang, D.: A 3D feature-based object recognition system for grasping. Master's thesis, School of Computer Science, University of Oklahoma, Norman, OK (2007)

[192] Wang, D., Watson, B.T., Fagg, A.H.: A switching control approach to haptic exploration for quality grasps. In: Proceedings of the Workshop on Robot Manipulation: Sensing and Adapting to the Real World at the 2007 Robotics: Science and Systems Conference (2007). Electronically published

[193] Webb, B.: Can robots make good models of biological behavior? Behav. Brain Sci pp. 1033–50 (2001)

[194] WHO: World Health Organization, International Classification of Functioning, Disability and Health (2001)

[195] Willett, R., Martin, A., Nowak, R.: Backcasting: Adaptive sampling for sensor networks. In: Proceedings of the third international symposium on Information processing in sensor networks, pp. 124–133 (2004)

[196] Woronowicz, M.: Selected afternoon constellation transient plume impingement model results. In: 3rd International Symposium on Formation Flying Missions & Technologies, Noordwijk, The Netherlands

[197] Xilinx, Inc.: Xilinx: Virtex-II Pro FPGAs. URL: http://www.xilinx.com/products/silicon_solutions/fpgas/virtex/virtex_ii_pro_fpgas/index.htm (2005)

[198] Yanofsky, P.: President wowwee robotics. personal communication (2005)

[199] Yeung, D.: Handling dimensionality and nonlinearity in connectionist learning. Ph.D. thesis, Department of Computer Science, University of Southern California, Los Angeles, California, USA (1989)

[200] Yeung, D., Bekey, G.: Using a context-sensitive learning network for robot arm control. In: Proceedings of the IEEE International Conference on Robotics and Automation, pp. 1441–1447 (1989)

[201] Yu, K., Chu, W.: Gaussian process models for link analysis and transfer learning. In: J. Platt, D. Koller, Y. Singer, S. Roweis (eds.) Advances in Neural Information Processing Systems 20, pp. 1657–1664. MIT Press, Cambridge, MA, USA (2008)

[202] Yu, K., Tresp, V., Schwaighofer, A.: Learning Gaussian processes from multiple tasks. In: Proceedings of the Twenty-Second International Conference on Machine Learning, pp. 1012–1019. Bonn, Germany (2005)

[203] Zhang, B., Sukhatme, G.S.: Adaptive sampling for estimaing a scalar field using a mobile robot and a sensor network. In: IEEE International Conference on Robotics and Automation, pp. 3673–3680 (2007)

[204] Zlot, R., Stentz, A., Dias, M.B., Thayer, S.: Multi-robot exploration controlled by a market economy. In: Proceedings of the IEEE International Conference on Robotics and Automation (2002)

# Index